DONALD DeYOUNG & DERRIK HOBBS

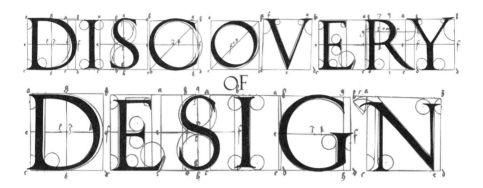

DISCOVERY OF DESIGN

Searching Out the Creator's Secrets

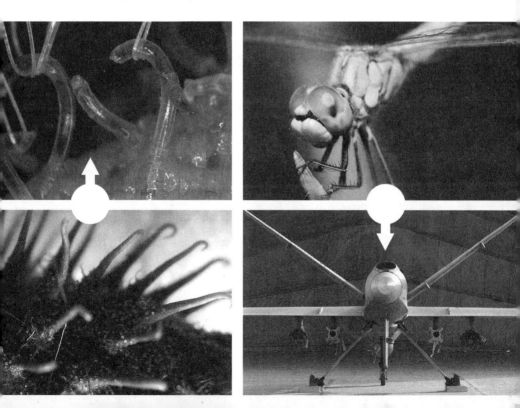

First printing: October 2009

Master Books®, P.O. Box 726, Green Forest, AR 72638.

ISBN-13: 978-0-89051-574-7
ISBN-10: 0-89051-574-3
Library of Congress Catalog Number: 2009930761

Cover Design: Thinkpen Design
Interior Design: Diana Bogardus

All Scripture quotations in this book, unless otherwise noted, are from the King James Version of the Bible.

Printed in the United States of America

Please visit our website for other great titles:
www.masterbooks.net

For information regarding author interviews,
please contact the publicity department at (870) 438-5288.

Master
Books®
A Division of New Leaf Publishing Group
www.masterbooks.net

ACKNOWLEDGMENTS

This *Discovery of Design* is dedicated to our wives, Sally DeYoung and Jessica Hobbs. We thank God for their faithful companionship and love.

— Don B. DeYoung and Derrik Hobbs

Cover images:

Velcro Macro
Tracy E. Anderson
Microscopist and Digital Imaging Specialist
Imaging Center, University of Minnesota

Cocklebur Hooks
Jim French
San Luis Obispo, CA

Dragonfly
Shutterstock.com

USAF Hunter-Killer Reaper UAV
James Gordon
US Department of Defense (DoD) Photographer

CONTENTS

Skin — Self-repairing Plastic
Tooth Enamel — Armor Coating
Vernix — Skin Cream

In crossing a heath, suppose I pitched my foot against a stone, and were asked how the stone came to be there: I might possibly answer, that for any thing I know to the contrary, it had lain there for ever: nor would it perhaps be very easy to show the absurdity of this answer. But suppose I had found a watch upon the ground, and it should be inquired how the watch happened to be in that place; I should hardly think of the answer which I had before given, that for any thing I knew, the watch might have always been there. Yet why should not this answer serve for the watch, as well as for the stone? Why is it not as admissable in the second case as in the first? For this reason, and for no other, viz., that when we come to inspect the watch, we perceive (what we could not discover in the stone) that its several parts are framed and put together for a purpose. . . . This mechanism being observed . . . the inference, we think, is inevitable, that the watch must have had a maker; that there must have existed, at some time, and at some place or other, an artificer or artificers, who formed it for the purpose which we find it actually to answer; who comprehended its construction, and designed its use.

— William Paley
Natural Theology, 1802

INTRODUCTION

Inventors and design engineers frequently look to nature for inspiration. There they find countless insights for new products and procedures. This book describes many of the useful results from this ongoing search. Nature is indeed a master teacher of design. And as a bonus, the products and designs found in nature arise from common, biodegradable materials. The name *biomimicry* is often given to this endeavor of discovering and utilizing designs from nature. Biomimicry and related words are defined in the glossary.

There are two distinct explanations for the host of successful ideas derived from nature studies. First, some people conclude that credit belongs to millions of years of evolutionary change. Over time, beneficial features in living things are said to be fine-tuned and optimized, while those organisms that are less fit are weeded out and eliminated. It is to be expected, some say, that exquisite designs are found throughout nature. After all, there have been millions of generations of trial and error to get it right. In this view, the brilliant tail of the peacock survives because earlier peacocks with short, drab tails failed in the competition to pass their genes on to later generations. There is, however, one major flaw with this natural explanation of design: it simply does not work. Patterns and information are conserved with the passing of generations, but the DNA blueprint does not increase in complexity or gain new information. A beautiful peacock tail does not gradually develop from fish scales, or from a knobby skin protrusion, or even from a short, drab tail. The occurrence of genetic mutations, including the occasional production of new species, actually displays an unavoidable loss or limitation of the earlier information content. Many scholars conclude that there is no convincing natural explanation for the peacock's tail or for any other design feature in living plants and animals.

There is a second explanation for the useful innovations found throughout nature. This alternative approach suggests a complete reversal of evolutionary progress over countless generations. It proposes that the valuable, practical design ideas surrounding us have been present from the very beginning of time. That is, useful features were embedded in the material universe by supernatural acts of creation. The purpose was for the benefit of living things, and also that ideas could be discovered and utilized for the welfare of mankind. In addition, design examples show us how to properly care for nature and maintain its health. Clearly, this explanation assumes intelligent planning by a beneficent Creator. Some might object that a divine hand in nature is not allowed. After all, today's science enterprise limits itself to naturalistic explanations for everything with no outside intervention. However, the historic definition of science is the search for knowledge and truth about the physical world, wherever this may lead. Regarding design in nature, the path of inquiry points directly to an intelligent plan.

Of special note is the book's eighth chapter regarding design found in nonliving parts of nature. Such items cannot somehow mutate or improve themselves over time. They have been present always. Also, chapter 7 describes some of the many medical benefits derived from plants and animals. Following each book entry there are questions for further study. Answers are provided at the end of the book. We also include a glossary of terms and a bibliography of biomimicry resources.

The authors of this book, along with many others, find the creation approach to origins and history to be a compelling and satisfying worldview. Readers are challenged to consider for themselves the alternate explanations for the limitless designs discovered in nature, and their implications. Explore new design at the website DiscoveryofDesign. com, and send us new examples and ideas for this growing database.

1
MICROORGANISMS

The microscopic world was explored in recent centuries by pioneer scientists such as Antonie van Leeuwenhoek (1632–1723). This Dutch pioneer built early microscopes and observed what he called "little animals." The world of microorganisms is crowded: living bacteria on your skin far outnumber the entire U.S. population. This small-scale life is neither primitive nor simple. Just the opposite. These tiny plants and animals reveal advanced designs for our study and benefit.

> In the year of 1657 I discovered very small living creatures in rain water.
>
> — Antonie van Leeuwenhoek

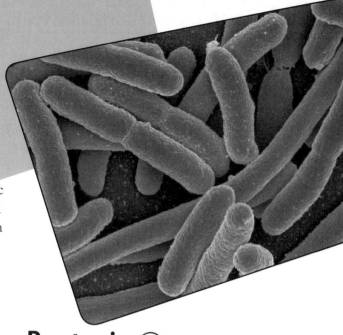

Bacteria ↶

Micro-motor

M any microscopic forms of life propel themselves through liquids using built-in protein assemblies called molecular motors. The organisms include certain bacteria and mitochondria, which exist within most living cells. The motor is in the form of a rapidly spinning filament called a flagellum, which functions much like a ship's propeller. A central shaft made of protein material spins as rapidly as 100,000 revolutions per minute (rpm), and is controlled by complex electrochemical reactions. These amazing "living motors" are able to stop and reverse their direction of turning in less than one rotation. Such flexibility is far beyond any manmade motor. Ten million of these molecular motors would fit along a one-inch length.

Cornell University researchers have succeeded in integrating molecular motors with metallic microspheres so that the bacteria transport the spheres through fluids. Future research goals include the use of the molecular machines as internal mobile pharmacies that deliver drugs exactly where needed within the body. *Discover* magazine describes these self-propelled bionic machines as one of the most promising emerging technologies.

As an alternative means of movement, consider the micrometer-size myxobacteria. This organism has hundreds of tiny nozzles covering its outer surface. It manufactures a slime that shoots from these nozzles, much like silly string. As a result, the bacterium recoils in the opposite direction using the principle of jet propulsion. The recoil speed exceeds ten micrometers per second. This is equivalent to a person traveling at 20 miles per hour, comparable to a swift runner. There are plans to duplicate this propulsion mechanism to control the movement of mechanical nanoscale devices within the human body.

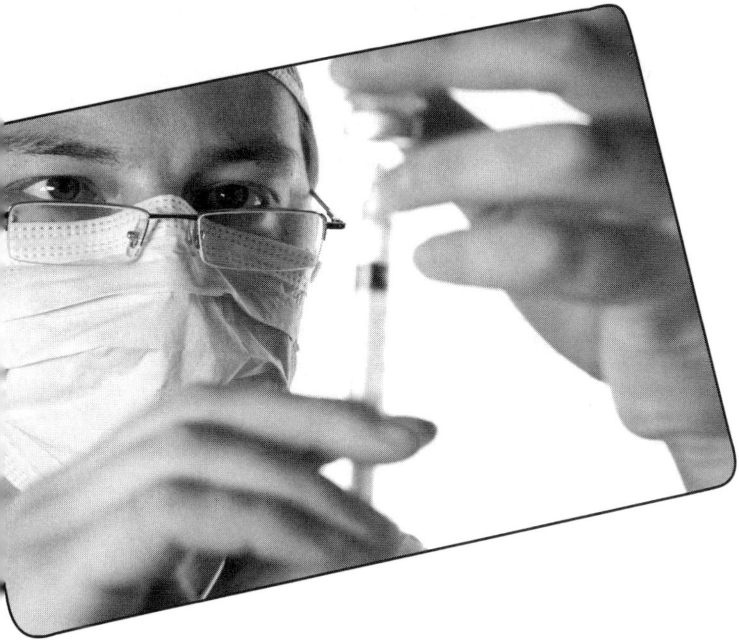

References

Goho, Alexandra. 2004. Mini motor. *Science News* 165(12):180.

Merali, Zeeya. April 1, 2006. Bacteria use slime jets to get around. *New Scientist* 192 (2545):15.

Questions for further study
1. What is the precise meaning of the words *micro* and *nano*?
2. How does the speed of an electric fan compare with the 100,000 rpm rate of the molecular motor?
3. What are the chemical properties of silly string?

A: pg. 188

Micro-motor

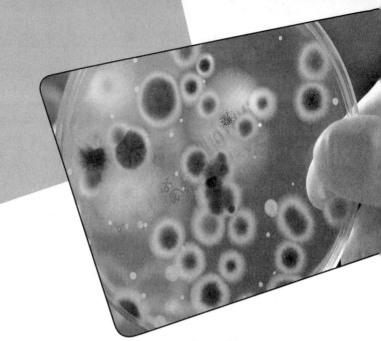

Bacteria ⟳ Battery

Scientists have taken a special interest in a bacterium called *Rhodoferax ferrireducens,* which resides in marine sediments. This tiny microbe produces electrical current using simple sugars as its fuel source. The bacterium feeds on the sugars, and a steady flow of freed electrons results. Waste materials are the bacteria's favorite diet. Sugars include fructose from fruit, xylose from wood, sucrose from sugar cane and beets, and glucose from many other sources. The electric energy production of the bacterium is more than 80 percent efficient, far above that of other organisms and man-made energy conversion processes.

Energy-producing microorganisms are known as bacterial batteries, or fuel cells. The technological challenge is to combine the electric output from a large number of these bacteria to produce a practical level of current. If successful, one cup of common sugar could light a 60-watt

bulb for many hours. This organic power source would be especially useful where the importing of fuel is difficult, such as remote villages. In such locations, the specialized bacteria could consume vegetation and turn the lights on.

Reference
Chadhuri, Swades, and Derek Loveley. 2003. Bacterial batteries. *Nature Biotechnology* 21(10):1229–1232.

Questions for further study

1. What actually is a battery?
2. Why are most energy conversion processes inefficient?
3. How many electrons pass through a standard 60-watt light bulb in one second?

A: pg. 189

Battery

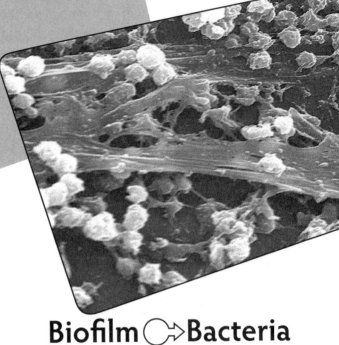

Biofilm ⟳➧Bacteria Control

B acteria are far from simple organisms. Some are able to organize themselves into large communities called *biofilms*. The bacteria residents communicate with each other by releasing chemicals.

On the microscopic level of bacteria, unseen chemical warfare goes on all around us. This cooperative activity of biofilms was first noticed on certain species of seaweed. In these underwater forests, friendly biofilms prevent the formation of foreign bacteria, which are harmful to the underwater plants. Studies have identified hundreds of chemical compounds, produced by biofilms, that affect their surroundings by "signal blocking."

One biofilm colony of special interest controls invading bacteria in animals. For example, cattle and dogs are susceptible to cholera infection, although they seldom become ill. The blocking biofilm prevents cholera bacteria from actively infecting the host animal. Our understanding of this benefit may permit the control of bacteria harmful to us by using "friendly" bacteria. Other specialized biofilms, when added to paint, prevent barnacles from attaching to boat surfaces. Further research

seeks to use biofilms to control corrosion in oil and gas pipelines. Such corrosion, often caused by bacterial growth, is a major problem in pipelines worldwide. Clearly, the potential medical and economic benefits of biofilms are enormous.

Internet search words:
biofilm, bacteria

Questions for further study

1. Estimate the number of bacteria on your hands.
2. Where might one find freshwater biofilms?
3. What are some unusual locations of biofilms?

A: pg. 189

Diatom

Nanotechnology

D iatoms are microscopic, single-celled algae. They are typically a few microns in diameter, ten times smaller than the width of a human hair. There are many thousands of distinct diatom species known, in both plant and animal varieties. They exist in countless numbers in the sea, freshwater lakes, and soil. They are the base of many food webs. Diatoms often house themselves inside intricate glass structures formed from silica or silicon dioxide, SiO_2. The intricate structures variously resemble stars, snowflakes, pyramids, chandeliers, cylinders, and crowns.

Scientists look to the diatoms for ready-made components in nanotechnology. One particular type of diatom resembles a circular glass gear, complete with an array of regular teeth around its outside edge. In the laboratory, this fragile structure is converted to a more durable form by heating at 900°C (1652°F) for several hours in the presence of magnesium gas vapor. The delicate silica glass vaporizes and is exactly replaced by the tough ceramic compound magnesium oxide with the chemical formula MgO. In an alternate process, diatom shapes can be converted to durable titanium oxide, TiO. Beyond the fabrication of microscopic mechanical gears, other diatom shapes are useful as tiny

Diatom

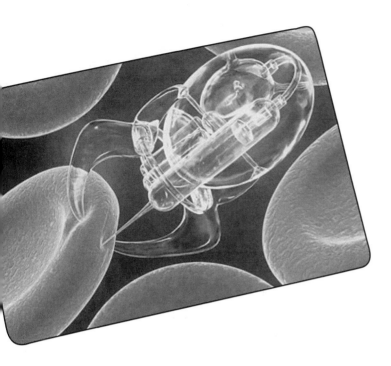

lenses. Still others with multiple pore openings can function as micro filters. Researchers also hope to coax diatoms to grow into new useful shapes called "designer diatoms." The common diatom displays master craftsmanship and unlimited applications.

References

Cohen, Philip. 2004. Natural glass. *New Scientist* 181(2430):26–29.
Goho, Alexandra. 2004. Diatom menagerie. *Science News* 116(3):42–44.

Questions for further study

1. Are diatoms plants or animals?
2. What is the mineral name for glass?
3. Diatomaceous earth is a powdered form of diatom fossils. What are some of its uses?

A: pg. 190

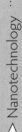

Protein ⟳ Solar Cells

Plants are very efficient at capturing energy from the sun by photosynthesis. The silicon solar cells manufactured today are far less efficient and tend to degrade over time.

Researchers have succeeded in fabricating small solar cells using photosynthetic proteins. The proteins are taken from plants and then deposited onto a glass surface. Thin, transparent layers of electricity–conducting material are also applied. When light shines on the proteins, they cause a faint current of electricity to pass through the adjacent layers. By placing multiple layers in series, the plant proteins generate

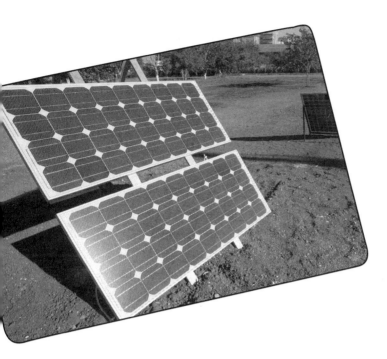

a useful electric current. One advantage of the protein solar energy cell is its ability to repair itself when deterioration occurs, since plants are self-healing. Solar energy is largely an untapped energy source, and plant proteins show us one way to proceed.

Reference

Das, R. and many others. 2004. Integration of photosynthetic protein molecular complexes in solid-state electronic devices. *Nano Letters* 4(6): 1079–1083.

Questions for further study

1. How is electric current measured?
2. How is it possible that wind power, water power, and fossil fuels are all forms of solar energy?
3. Can you name three nonsolar forms of energy?

A: pg. 190

Solar Cells

2

THE INSECT WORLD

It is estimated that 90 percent of distinct living animals on earth remain unknown to science. Many of these undiscovered creatures live in tropical rain forest regions, and most are insects. The word *insect* derives from the Latin for "something cut in," describing the insect's segmented body. This large group of invertebrates includes ants, beetles, and flies. The world of insects is often popularly expanded to include other small creatures, such as spiders and centipedes. The following examples show the important lessons that insects teach us.

Else, if thou refuse to let my people go, behold, tomorrow will I bring the locusts into thy coast .

— Exodus 10:4

Many of us have watched a large flock of birds circling in the sky with ever-changing patterns. How do they keep from colliding, and what determines the group's direction of travel? This complex, organized behavior is called "swarming." It occurs for creatures in the air, on the ground, and under water. Ants, birds, fish, fireflies, honeybees, and even herds of caribou display swarming behavior.

Ants ↻→ Airlines

The swarming or foraging behavior of ants has been studied in detail. It is found that no single ant is in charge of an entire nest. Instead, countless interactions between nearby ants determine the overall group behavior. Contact between their antennae, and shared smells, give information on the home base of the ants, nest repair needs, food sources, predators, and much more.

The self-organizing ability of creatures is modeled with computer programs that guide us in problem solving. One example concerns the efficient management of major airports. Continual decisions are required for the order of aircraft takeoffs and landings, runway use, and proper gates. The manager's goal is to minimize the travel and waiting times for passengers. Input data to the computer programs includes the time needed for various aircraft, large and small, to get into and away from each gate. Other time factors include luggage transfer, refueling, and

aircraft cleaning between flights. The computer output suggests the schedule for safe, optimum airport operation with minimum delays.

Similar ant-imitating computer programs are finding widespread application wherever complex scheduling exists. Examples include the management of telephone networks, mail distribution, computer search engines, truck fleets, and even boardroom decisions. Thanks are due to the ant world for helping to solve traffic problems!

Reference

Miller, Pete. 2007. Swarm theory. *National Geographic* 212(1):126–147.

Questions for further study

1. How many legs does an ant have?
2. How does the total number of ants compare with other animals?
3. What did King Solomon say about ants?

A: pg. 191

Asian Beetle ⟳⇢ Paper Whitener

Several species of beetles in Southeast Asia have a brilliant white appearance. Their surface color is whiter than common substances such as milk or tooth enamel. Instead of a pigment, the extreme whiteness of the beetle results from tiny, translucent surface scales made of chitin, a common material in the natural world. The structure of these scales consists of fiber filaments, just 250 nanometers in diameter. The random orientation of these fibers causes incoming light to reflect in all directions and at all visible wavelengths. The resulting color, pure white, is similar to that reflected from a layer of fresh snow.

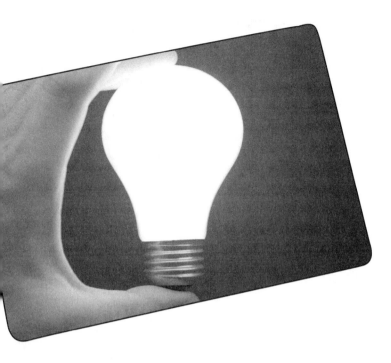

There are several possible applications based on the white *Cyphochilus* genus of beetles. Artificial light-scattering filaments could replace the tons of minerals used to brighten paper. The fiber material could also coat the inside of lightbulbs. This would make lights more energy efficient, and the output would more closely match the familiar sunshine.

Reference

Perkins, Sid. 2007. Micro-structures make a beetle brilliant. *Science News* 171(5):78.

Questions for further study

1. What is currently used to give paper a white color?
2. What is chitin?
3. What could be the purpose of the Cyphochilus beetle's white appearance?

A: pg. 191

Paper Whitener

Asian Beetle

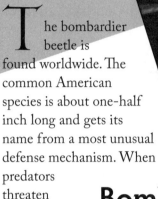

T he bombardier beetle is found worldwide. The common American species is about one-half inch long and gets its name from a most unusual defense mechanism. When predators threaten the beetle, it produces repeated chemical explosions using

Bombardier ⟳→ Gas Turbine Beetle Engine

a tiny combustion chamber within its abdomen. The result is a pulsed jet of hot, vaporized fluid, fired outward from a nozzle and aimed in almost any direction. The temperature has been measured at about 100°C (212°F), equal to that of boiling water. The jet of mostly steam travels at a remarkable 65 feet (20 m) per second. One propellant in the spray is hydrogen peroxide (H_2O_2), a potent chemical that is useful as rocket fuel. Frogs, spiders, ants, and birds learn to keep their distance from bombardier beetles after an unpleasant scalding by the irritating fluid. A distinct popping noise also confuses any would-be predators.

The bombardier beetle may help solve a particular problem that occurs during aircraft flight. Gas turbine engines sometimes quit during flight and must be reignited quickly. When this emergency occurs, an electrode fires and sends a stream of charged particles into the engine to restart it. This procedure is sometimes successful and at other times fails to restart the engine. The bombardier beetle's combustion process provides useful designs for the aircraft engines. These include the beetle's heart-shaped combustion chamber, and also the rapid rate of firing that produces 500–1000 pulses of jetted steam per second.

Beyond aircraft engine design, the bombardier's defense technique may lead to improvements in rocket technology, and also the emergency inflation of automobile air bags.

In 2007 researchers at Oregon State University identified a fossil with behavior similar to the bombardier beetle. A soldier beetle was found entombed in amber while in the act of spraying a chemical propellant at an unknown attacker. It is thought that this amber dates to the time of the dinosaurs. Clearly, beetles have displayed impressive defense mechanisms ever since their creation.

References

Dean, J., Aneshansley, D. J., Edgerton, H.E. and T. Eisner. 1990. Defensive spray of the Bombardier Beetle: a biological pulse jet. *Science* 248(4960):1219–1221.

Ross, Greg. 2004. Insect Inspiration. *American Scientist* 92(2):128–129.

Questions for further study

1. Hydrogen peroxide is found in many medicine cabinets. What is its use?
2. Where are bombardier beetles found?
3. How would an evolutionist attempt to explain the origin of the bombardier beetle?

A: pg. 191

Gas Turbine Engine

Butterfly ⟳ Cosmetics

Many butterflies have wings that shimmer with brilliant colors in sunlight. For example, the *Morpho rhetenor* (MOR-foh REH-teh-nor) butterfly wing is an iridescent blue. More than 80 distinct species of the colorful Morpho butterfly live in South America. Their bright color arises from thousands of tiny overlapping scales that make up the wings. As sunlight penetrates these layers, only the blue color reflects back. That is, the light wavelengths of the color blue blend, while other color wavelengths cancel out. Scientists describe this as the constructive and destructive interference of light waves.

Cosmetics are available that mimic the beauty of butterfly wings. The company L'Oréal has copied the butterfly design by inserting tiny, layered flecks of either mica or silica into eye shadow, lipstick, nail polish,

and mascara. The spacing between the internal layers of the mineral flecks determines which particular color reflects back. This "butterfly effect" replaces the standard use of pigments, waxes, and oils to produce various cosmetic colors. The iridescent butterfly effect is also used for car paint, credit card security, and optical computing.

Questions for further study

1. How many different species of butterflies have been catalogued?
2. How long does a butterfly live?
3. What causes the different colors of light?

Reference

Cunningham, Aimee. 2005. Way to glow. *Science News* 168(21):324.

A: pg. 192

Dragonfly

Surveillance

Dragonflies often fly above wetlands and ponds in search of food. The origin of their name is uncertain. Dragonflies do not bite or sting people, and contrary to some traditions, they do not sew together the lips or eyelids of bystanders! Instead, the dragonfly is an outstanding and useful example of design.

Dragonflies are unusual insects in that they have four distinct wings. Up-and-down strokes of these wings allow the insect to hover motionlessly or to shift quickly into forward or reverse motion. The forward speed of dragonflies reaches nearly 40 mph (64 km/hr). On each downward stroke, a dragonfly twists its wings slightly. This causes a whirl of air above the wing and reduces the air pressure slightly, resulting in the needed lift for flight.

Dragonflies pursue their prey with maneuvers that are far beyond the capability of jet fighters. By flying in a direct line with prey, dragonflies give the visual illusion of not moving at all. This disguise of apparent

motionlessness is highly successful in the
hunting of smaller insects. Military strategists
hope to duplicate the flying ability of the dragonfly. One possible
application is a small, unmanned drone or surveillance aircraft that
could hover motionless, then dart suddenly to a new location just as
dragonflies do. On a larger scale, helicopters could benefit from increased
maneuverability copied from dragonflies. Dragonflies are able to carry
as much as 15 times their own weight while in flight. Modern aircraft
cannot carry a load much heavier than their own weight. Clearly,
dragonflies have much to teach us about successful flight.

Reference

Graham-Rowe, D. 2003.
You'll never see it com-
ing... *New Scientist*
178(2401):18–19.

Questions for further study

1. What becomes of dragonflies in
   wintertime?
2. Have fossil dragonflies been
   found?
3. What do dragonflies eat?

A: pg. 192

Surveillance

A firefly produces its "cold light" by mixing chemicals in its abdomen. Oxygen is combined with a substance called luciferin in the presence of the enzyme luciferase, which catalyzes or activates the

Firefly ↻→ Light Stick

reaction. Energy from this oxidizing reaction excites electrons, and they in turn emit visible light. This light-producing process by living things is called bioluminescence. The insects synthesize the needed chemicals from their food supply. No batteries are required for firefly flashlights.

Firefly light was first duplicated in the 1960s by chemical engineers at Bell Labs and also at American Cyanomid. The artificial process is called chemiluminesence. Research continues today to produce glowing colors with various chemical dyes. Green and yellow colors are relatively easy to make, while red and purple are more challenging.

Commercial light sticks keep the two component chemicals separate inside a plastic tube. Bending the tube then breaks a small internal glass vial and mixes the chemicals. The glowing light turns on immediately and may last for several hours. Light sticks are used for emergency signals, wilderness camping, and by commercial fishermen as markers for nets and lines.

There are more than 2,000 known species of firefly. Their eggs produce a soft green-white light, as do the larvae, called glowworms.

The firefly patterns of flashing light are used to communicate and identify mates. Bioluminesence is also very common in marine life. Almost all deep-dwelling species produce chemical light of varied color. In some cases, luminescent bacteria live within the marine creatures, causing them to glow from the inside out.

Reference

Marchant, Joanna. 2000. First light. *New Scientist* 167(2248):34–35.

Questions for further study

1. What is the lifespan of the firefly?
2. What is the origin of the chemical names luciferin and luciferase?
3. Why does a group of fireflies sometimes blink off and on simultaneously?

A: pg. 193

Light Stick

Fly ↻→Hearing Aid

Hearing aids have a number of shortcomings. Some wearers claim discomfort, and also an inability to follow a conversation in a noisy room. Most receivers pick up background sounds and produce noise in the ear. A small fly, *Ormia ochracea*, may provide a solution. This fly has the incredible ability to detect and track individual sounds. Living in a large open field, the female fly tracks the chirping sound of a particular species of cricket. Locating the source of sound, the fly then deposits eggs on the back of the singing cricket.

We do not know how the fly is able to focus on an individual cricket in the vast outdoor world of sounds. For people, the direction to a sound source is noticeable because pressure waves reach one eardrum an instant before the other. The slight difference in arrival time at our two ears, about 0.0004 seconds, is processed by our brains to determine direction. The hearing mechanism of the fly, however, is tiny in comparison.

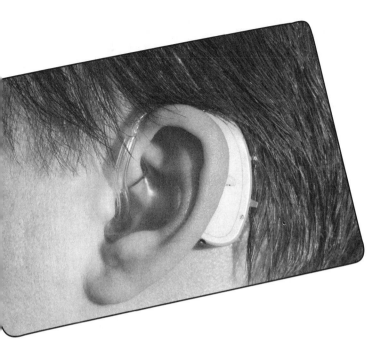

There cannot be a discernable difference in arrival times between insect ears. Instead, there must be an alternate mechanism used to determine the precise direction of the cricket chirp. Hearing aids using this yet-undiscovered technique may allow wearers to focus on just one person's speech. As an added benefit, a functional fly-sized hearing aid would be nearly invisible.

Internet search words: fly, hearing aid

Questions for further study

1. How may a cricket be used as a thermometer?
2. How does hearing sensitivity compare with our other senses?
3. Does a fly have eardrums?

A: pg. 193

Hearing Aid

Honeybee ↻
Surveillance

Honeybees, flies, and many other insects have compound eyes. Each eye consists of thousands of tiny tubes or columns called *ommatidia*. Light first enters through a micro lens, which caps the outer end of the column. The light then moves through the hollow column until it meets photoreceptors at the internal end. Insect eyes have the ability to see both lighter and darker images at the same time. Cameras typically are limited to either light or dark illumination, but not both.

Berkeley scientists are working on an artificial eye that duplicates the advanced optics of the honeybee. They begin with a tiny lump of clear resin. Thousands of tiny bumps are formed on the surface to function as lenses. Next, the resin is exposed to ultraviolet light. The surface lenses divide this light into many separate beams that move through the resin. Along the way the light polymerizes or chemically

alters the resin. The result is permanent, side-by-side columns of light-guides within the resin, similar to the ommatidia of insects. The ongoing laboratory challenge is to lengthen the light guides and also to connect them with microelectronic photo sensors. Applications include miniature surveillance cameras and medical endoscopes for probing inside the body.

Reference

Weiss, Peter. 2006. Rounding out an insect-eye view. *Science News* 169(20):318.

Questions for further study

1. What image does an insect see with its compound eyes?
2. How small is one of the multiple lenses in the honeybee eye?
3. What is ultraviolet light?

A: pg. 194

Surveillance

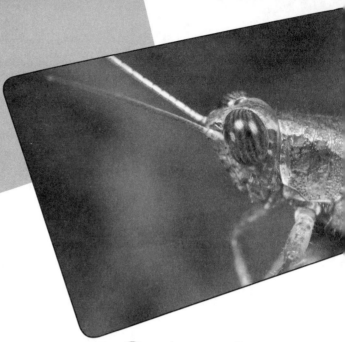

Insect ↻ Atomic Hearing Force Microscope

The Atomic Force Microscope, or AFM, obtains images of samples on the smallest scale of present technology. Similar to the needle on an old-fashioned record player, the AFM has an arm with a sensitive, pointed probe. As this probe moves directly above a sample, it records the changing electric field caused by minute surface details. This data then is transformed into an enlarged, three-dimensional image of the surface. One ongoing challenge is that the AFM probe itself sometimes alters the delicate surface that it is exploring.

Insects such as grasshoppers and moths have listening systems that are ultra-sensitive to very faint vibrations. This helps them locate food and avoid nearby predators. Scientists are using the Atomic Force Microscope to find out how insects are able to "see" and "hear" vibrations

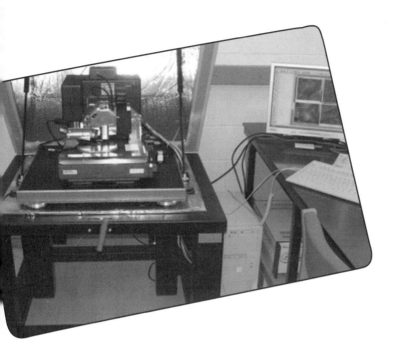

without actually touching or disturbing the object they are exploring. In turn, this information will lead to improvements of the microscope itself. The goal is the study of delicate biological materials without altering them in the process.

Internet search words:
  Atomic Force Micro-
  scope, Bristol University,
  insect

Questions for further study

1. Who invented the Atomic Force Microscope?
2. How small an object can an AFM clearly "see"?
3. List some of the different types of microscopes.

A: pg. 194

Insect Hearing ← 41

Insects ◯→Robotics

There are two major competing ideas for the movement of small robots: wheels and mechanical legs. Robotic planet probes sent to Mars, for example, have multiple wheels. However, these remotely operated vehicles become easily stuck in sand. In 2005 the Mars probe *Opportunity* was stuck in sand for five weeks. Scientists remotely rocked the probe back and forth until it was finally free. Meanwhile, in the natural world, legs provide efficient locomotion. The detailed movement of insect legs is studied at the University of California at Berkeley. The school operates the PEDAL Laboratory, standing for the performance, energetics, and dynamics of animal locomotion. Insect-robotics lab research is also carried on at Stanford, Harvard, and Johns Hopkins Universities.

At the Berkeley PEDAL lab, cockroaches and centipedes are filmed as they run on miniature treadmills. Tiny "jet packs" are attached to the insects to record their reaction when they are thrown off balance by sideways spurts of air. The insects are very talented at maintaining their balance, even while running on a rough surface. The insect legs typically function like tiny pogo sticks with a bouncing stride.

Researchers at the PEDAL lab are convinced that legs are the optimum way for small robots to navigate. Models with six or eight legs have been constructed, based on the mechanics of insects. These models

have one major advantage over insects: If the robot is overturned, the mechanical legs can be pivoted 180 degrees downward and the upside-down machine can continue walking forward. Applications with robots include the exploration of planet surfaces with irregular terrain, the detection of explosives, cleanup of toxic spills, and similar hazardous tasks.

References

Burdic, Alan. 2004. Building the perfect pest. *Discover* 25(7): 24-25.

Dartnell, Lewis. 2005. Creepy crawlies to explore other worlds. *New Scientist* 187(2509):24-25.

Questions for further study

1. What is the origin of the word *robot*?
2. Can one obtain an advanced degree in robotics?
3. How many legs do centipedes have?

A: pg. 195

Robotics

Namib ⟳ Water Beetle Collector

The Namib (nä´mib) beetle, also called the African Stenocara beetle, lives in the deserts of Namibia, South Africa. The very name *Namib* means "an area that is desolate." The sparse rainfall in this region is less than one inch per year. There is frequent morning fog, however, and the beetle knows how to wring moisture from the air. Its bumpy outer wings collect and combine the tiny fog droplets. Upon reaching a certain size, the water droplets overcome electrostatic attraction forces and roll down the beetle's tilted back to its mouth. The Namib beetle's back has bumpy hydrophilic (water-attracting) portions that alternate with hydrophobic (water-repelling) channels or grooves. The droplets coalesce and grow around the bumps, and then flow along the grooves.

Scientists have developed a synthetic surface that mimics the beetle's efficient water-gathering technique. Small, poppy seed-size spheres are embedded in a thin layer of wax. When exposed to fog, this layer readily attracts moisture. Tests show that neatly ordered arrays of bumps, similar to the beetle's back structure, catch maximum mist from the air. Such materials can provide a source of water in a desert region whenever fog or dew appears.

In Israel, an alternate technique is used to harvest moisture. Perhaps you have seen morning dew droplets glistening on a spider web. This has inspired the invention of a large web of netting called *WatAir*, resembling a tent turned upside-down. At night, droplets collect on the web and roll downward into a bucket placed beneath the center portion. A web 20 feet across collects up to 12 gallons of water by morning.

References

Fiscketti, Mark. 2006. Call it beetle guard. *Scientific American* 295(6):33–34.

Summers, Adam. 2004. Like water off a beetle's back. *Natural History* 113(1):26–27.

Questions for further study

1. What causes fog or dew to appear in the morning?
2. What defines a desert climate?
3. How do Namib beetles communicate?

A: **pg. 195**

Water Collector

Spider thread can be as thin as 10 nanometers, 10,000 times thinner than a human hair. Spiders use these fine strands to build their webs, wrap prey, and occasionally to "ride the wind" to a new location.

Spider ⟳ Fiber
Silk   Optics

Material engineers at the University of California have utilized spider silk to make ultra-small hollow tubes. First, they dip the silk repeatedly into a chemical solution of tetraethyl orthosilicate, $Si(OC_2H_5)_4$. After the coated fibers dry, they are baked at 788°F (420°C), which burns away the interior silk. The coating also shrinks fivefold and forms a flexible, hollow tube made of silica, or glass, now 50,000 times thinner than a hair follicle.

The resulting "nanotubes" have at least three important applications. First, they readily conduct light as a tiny version of fiber optics. These "light pipes" are useful for ultrafast nanoscale optical circuits. Second, the hollow fibers provide microscopic test tubes that can hold single molecules of specific chemicals. These tubes can function as sensitive chemical detectors. They also provide an opportunity to study the unusual behavior of very small assemblies of molecules. As a third application, the nanotubes find use in specialized microscopes that provide extreme magnification.

It is somewhat of a mystery how spiders are able to manufacture their fine silk, using chiefly air and water as raw products. This silk is

five times stronger than an equivalent weight-for-weight strand of steel. It is also unknown how spiders prevent the silk from clogging their spinnerets where the strands are secreted from the body. The polymer threads solidify instantly when the chemical is exposed to the air. Spiders clearly display advanced engineering design, and their silk output is a valuable resource. Synthetic spider silk is one of the most sought-after technologies in biomimicry. Possible products range from bulletproof vests to suspension cables for bridges. Thus far, success in duplicating spider silk production has been limited.

Reference

Penman, Danny. 2003. Spiders weave a web of light. *New Scientist* 117(2387):20.

Questions for further study

1. What was an early use of spider silk by astronomers?
2. What is a nanotube?
3. How can it be said that spider silk is stronger than steel?

A: **pg. 196**

Fiber Optics

Termite ⟳Ventilation Mound

Termite mounds across the plains of southern Africa reach heights of 10 feet (3 m) or more. Engineers have long been impressed with the self-cooling system built into the mounds by *Macrotermes michaelseni* termites. Their food supply is a "farmed" fungus that must be kept at exactly 87°F (30°C). However, the outside temperature varies widely between 35 and 104°F (2–40°C). To compensate, the termites open or close a series of internal heating and cooling vents. These vents connect numerous tunnels that maintain temperature along with ideal moisture and oxygen levels. The network of tunnels somewhat resembles our internal circulatory system of veins and arteries. When rain occurs, clay on the surface of the termite mound swells and provides waterproof protection. During dry periods, the clay contracts and ventilation cracks appear.

Architects are making use of the principles of termite mound ventilation. One result is the Eastgate Building in the capital city of

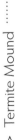

Zimbabwe, Harare. This 18-story shopping complex has no mechanical air conditioning or heating, yet remains comfortable. The interior ventilation system of ductwork is based on the structure of termite mounds. Outside breezes pull fresh air throughout the building. As a result, the Eastgate structure uses just 10 percent of the energy of a conventional building.

Internet search words: air conditioning, Eastgate Building, termite

Questions for further study

1. How many termites may live in a single African mound?
2. Describe how the termite mound heating and cooling vents operate.
3. What is the average temperature of Zimbabwe's capital city, Harare?

A: pg. 196

Ventilation

Timber ⟳→Chainsaw Beetle Larva

G as-powered chainsaws were first developed a century ago for harvesting timber. The early saw chains had pointed teeth shaped similar to those on manual hand saws. However, these teeth required frequent sharpening and maintenance.

One autumn day in 1946, Oregon logger and inventor Joseph Buford Cox was chopping wood. During a rest break he noticed the activity of timber beetle larvae, also called wood sawyer. These thumb-size insects were steadily chewing their way through a tree stump, some along the wood grain and others against the grain. Close inspection revealed small C-shaped jaws or mandibles on the beetles that sliced cleanly through the wood fiber. The pair of sharp mandibles moved sideways in a pinching, scissors fashion. During the next year, Cox designed and built

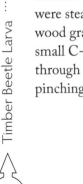

the "chipper chain," which is still widely used today. This chain has teeth with curved, C-shaped cutting edges, similar to the beetle's jaw. The teeth cut much more efficiently than earlier saw designs. Cox's original Oregon Saw Chain Company today has grown into a major forestry industry. Many chainsaws purchased today have an Oregon chain.

Internet search words: chainsaw, Joseph Cox, Oregon chain

Questions for further study

1. How were big trees harvested before chainsaws were invented?
2. Are timber beetles a major pest in forestry?
3. What is the largest commercial chainsaw manufactured today?

A: pg. 196

Chainsaw

Wasp ⟳→ Paper

In centuries gone by, paper was made from cotton and linen rags. The resulting paper was expensive and limited in quantity. Around 1719, French scientist René-Antoine Réaumur noticed the parchment-like nests of paper wasps, made from vegetation. Réaumur wrote, "The American wasps form very fine paper.... They teach us that paper can be made from the fibers of plants without the use of rags and linens, and seem to invite us to try whether we cannot make a fine and good paper from the use of certain woods."

In modern paper mills, cellulose fibers in wood are separated from their lignin binder. Water, filler, and various new binders are then added to make wood pulp. Pressing and drying of the mixture results in a great variety of paper, ranging from tissue to currency. The humble paper wasp has led the way in paper production.

Internet search words: paper, Réaumur, wasp

Questions for further study

1. What is the shape of the paper wasp nest?
2. What is lignin?
3. Is some paper still made from linen cloth?

A: pg. 197

Water ⟳→ Water
Strider   Repellant

I n chapter 5 we describe brush-like fibers on the feet of the gecko lizard. This feature allows the creature to walk upside-down across ceilings. The brush feature also occurs on the legs of the water strider.

Perhaps you have watched these slender insects skim easily across the surface of a pond, as if they are skating. They easily "walk on water" while staying completely dry. It was earlier thought that a waxy coating on the strider's feet repelled water. A microscopic view, however, reveals thousands of tiny hairs on the strider feet, called micro setae. Air is trapped in the tiny spaces between the hairs, and the result is a water-repellent barrier. Researchers in China have measured the extent to which water striders can avoid submersion. Scientists made an artificial strider leg and pressed it against a film of water. The leg made a depression in the water layer but easily supported 15 times the insect's weight before penetrating the water surface.

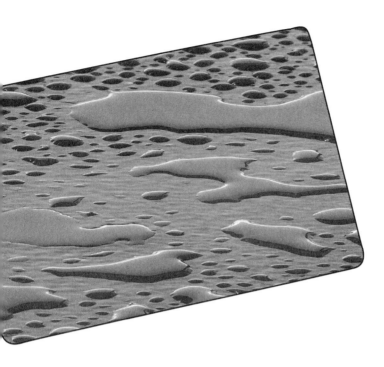

The microscopic hairs on the water strider are present all along the insect's legs. This ensures that the strider remains safely on a water surface without drowning, even during a heavy rain. The design of the micro setae hairs may lead to miniature aquatic robots and other non-wetting materials.

Reference

Gao, Xuefeng, and Rei Jiang. 2004. Water-repellent legs of water striders. *Nature* 432(7013):36.

Questions for further study

1. How do water striders move from one pond or puddle to another?
2. How is it that a water strider, or even a needle, can float on water?
3. How do water striders move so rapidly on water?

A: **pg. 197**

Water Repellant

3

FLIGHT

Flight surely stands near the top in our amazement with living creatures. Insects, flying mammals, and birds display acrobatics far beyond the ability of modern aircraft. And regarding annual migrations, we still have much to learn about the timing and directional abilities of birds. The following examples give an introduction to flight lessons from winged creatures.

> The flowers appear on the earth; the time of the singing of birds is come, and the voice of the turtle is heard in our land.
>
> — Song of Solomon 2:12

Bats display an amazing ability called echolocation. Their vocal cords emit sound waves with frequencies in the range 20,000 to 120,000 cycles per second, also called *hertz*. Human hearing has a much lower and more limited frequency range, between 20 and 20,000 cycles per second. Bats also have the ability to

Bats ⟳ Sonar Systems

detect the reflection of their emitted sound waves from nearby objects. Experiments show that bats can sort out distinct echoes that arrive just two or three microseconds apart. This amounts to a distance variation that is less than the width of a hair. With such precision, bats are able to mentally construct three-dimensional images of objects. They also fly safely in total darkness and capture their prey while avoiding obstacles such as tree branches.

Bats use a strategy called *parallel navigation* to catch insects. In this technique, the bat keeps the compass direction to the target a constant by changing its flight pattern as needed. If a moving insect is initially located in the northeast, for example, the bat maneuvers to always keep the insect in the same northeast direction. During flight, the head of the bat also maintains a constant orientation or tilt. In contrast, a baseball fielder uses the *constant bearing* method by following the ball with the eye and turning the head upward. In the 1940s, when guided missiles were developed, engineers realized that parallel navigation is by far the best way to hit the target. Bats had this ability all along.

In the sea, dolphins, and perhaps whales, also navigate by using echolocation. The detection ability of these animals remains far beyond the current capability of modern electronics. The U.S. military conducts ongoing studies of bats and dolphins to improve sonar systems. The word *sonar* comes from **SO**und **N**avigation **A**nd **R**anging.

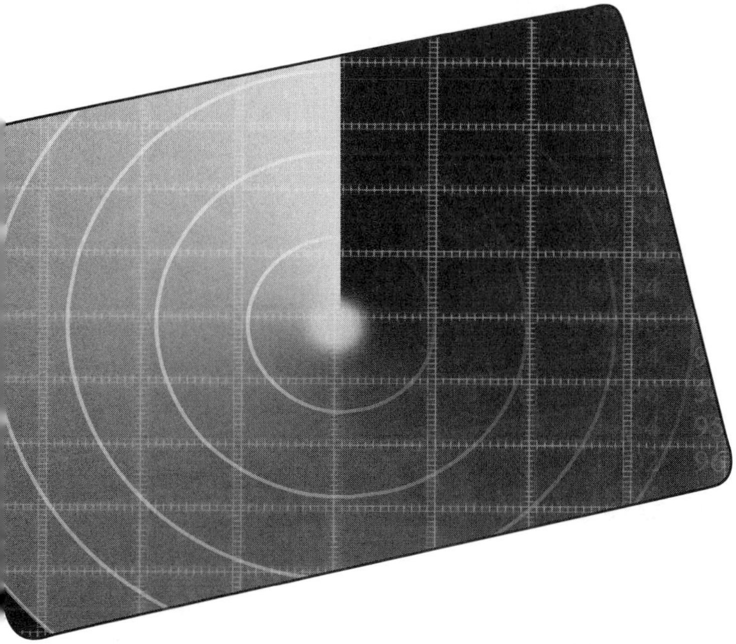

To aid the blind, a bat-inspired sonar walking stick is available. This device emits a signal of 60,000 hertz, far above our hearing range. The return echo from nearby obstacles is detected by an electronic sensor and then converted into vibrations of several buttons built into the walking stick. The blind person is made aware of the nearness of objects by sensing the button vibrations. The buttons also provide information on the direction and size of obstacles.

Reference

Perkins, Sid. 2005. Learning to listen. *Science News* 167(20):314–316.

Questions for further study

1. What actually is a sound wave, as produced by bats?
2. What is the frequency of a dog whistle?
3. What is a flying fox?

A: pg. 198

Sonar Systems

Bird Flight ⟳→Aircraft

Studies of birds show that they use minimal effort, or energy, during flight. Birds typically cruise at a speed that allows them to slip most easily through the air. Their wings and tail feathers can produce turbulence in the air called *eddies*. If a bird flies too fast, its wings must continually fight the drag that results from the swirling air. In contrast, if the flight is too slow, air turbulence tends to "stick" to the feathers and interferes with the forward motion.

There is a math formula that describes bird flight. The maximum stroke frequency of the wing is multiplied by the vertical distance traveled by the wing tip during the flapping stroke. This number is then divided by the bird's forward speed. The result is a number called the *Strouhal* number. A value in the range 0.2–0.4 is considered to be efficient flight. With some minor alteration, this same formula also applies to

the swimming motion of fish and whales. Engineers use the flight-swim formula to design small aircraft for the military, as well as underwater submersibles. The successful animal world instructs us regarding the proper speeds for efficient flight or water travel.

Reference

Taylor, G.K., R.L. Nudds, and L.R. Thomas. 2003. Flying and swimming animals cruise at a Strouhal number tuned for high power efficiency. *Nature* 425(6959):707–711.

Questions for further study

1. What are the fastest speeds for animals in the air and under water?
2. Give a numerical example for the bird flight formula.
3. Do flying fish have wings?

A: pg. 198

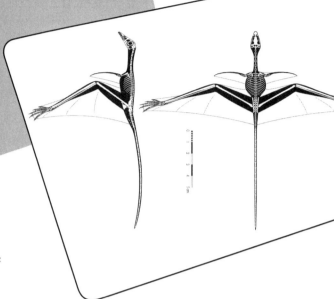

A single fossil of the flying reptile *Sharovipteryx mirabilis* was found in Kyrgyzstan some years ago. The creature was about eight inches long (20 cm), and had an unusual wing structure. Today's flying vertebrates,

Flying ⟳→ Delta Wing Reptile Aircraft

including birds and flying squirrels, have wings or membranes attached to their front limbs. Reconstruction of the fossil, however, shows the wings attached to the hind limbs in a delta-wing pattern. Wind tunnel experiments show that this design gives excellent flight performance. The rear-limb design outperforms the ability of today's gliding lizards. As a side benefit, the rear wing would not interfere with walking since the wing membrane was elastic and could be folded against the reptile's body.

Delta-wing aircraft, including the Eurofighter Typhoon jet, are designed similar to the fossil lizard. The delta wing allows high-speed performance while maintaining maneuverability. Both the fossil lizard and modern aircraft also have small triangular wing structures up front. On aircraft, this forward wing is called the "canard," and it allows slower and safer landing speeds.

A question arises for evolution theory: How is it that a flying creature from the past has several design advantages above and beyond most modern creatures? Instead of evolutionary progress, the fossil lizard indicates the loss of a particular flight design.

Flying Reptile

Five centuries ago, Leonardo da Vinci (1452–1519) designed "flying machines" based on the flapping wings of birds. His early attempts at flying were not successful. The Wright brothers later realized that larger birds tend to glide, rather than constantly flapping their wings. The brothers studied many birds, including vultures, to learn the details of bird flight. This led to the successful fixed wing structure for the Wright brothers' airplanes.

References

King, Anthony. 2006. The graceful glide of a delta-winged lizard. *New Scientist* 109(2545):16.

Shipman, Pat. 2008. Freed to fly again. *American Scientist* 96(1):20–23.

Questions for further study

1. What is the meaning of the name *Sharovipteryx mirabilis* given to the fossil flying reptile?
2. What is a wind tunnel?
3. What year did the Wright brothers succeed with their flight?

A: pg. 199

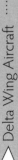

Delta Wing Aircraft

S ome of
the fastest
electric trains
in the world are
operated in Japan.
They connect many
cities and travel in
excess of 200 miles
per hour (322 km/
hr). The trains have
an excellent safety
record; however,
there are new problems
to solve. Many tunnels are a part of the Japanese rail system. When a
train exits a tunnel at high speeds, there is a rapid expansion of air that
was compressed in front of the train. This results in a loud sonic boom
that rattles windows and awakens people. Japan has strict laws on sound
pollution, so design engineers went to work and found the solution
in nature.

Kingfisher ⟳ Bullet Train

The kingfisher bird looks comical with its stubby tail, large head, and
long, sharp beak. However, this bird is an excellent fisherman, diving
straight down into water. It is noticed that when the kingfisher enters
water at high speed, there is very little splash. The kingfisher is the envy
of Olympic divers. When the bird hits the water, it experiences a drastic
change in pressure. This difference is somewhat similar to the air pressure
change of a bullet train emerging from a tunnel into open air. Wind
tunnel experiments show that the kingfisher's bill is ideally shaped for a
smooth, streamlined transition from air into water. The fronts of many
Japanese bullet trains have now been redesigned to mimic the shape of
the kingfisher's bill. Many train engines are given long, futuristic, tapered
noses. As a result, sound is greatly diminished when exiting tunnels, and

Kingfisher

overall vibration is less. As a bonus, total energy consumption is reduced by 15 percent. Even the recessed headlights of some trains are modeled after the nostrils of the kingfisher. This bird leads the way in achieving smooth, quiet train operation. As biomimicry expert Janine Benyus remarks concerning the kingfisher, "We are surrounded by genius."

Questions for further study

1. Where do kingfishers live?
2. How do kingfishers see while under water?
3. What is a sonic boom caused by the bullet trains?

Reference

Benyus, Janine. 1997. *Biomimicry: Innovation Inspired by Nature.* Harper Collins Publishers, New York.

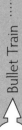

A: **pg. 199**

Kingfisher 65

Bullet Train

Owl Wing ⟳⇢Noise Reduction

Owls are one of the quietest flying birds. Their silence allows them to catch prey by complete surprise. Two special design features make their stealth flight possible. First, the forward fringe on the owl's wing is uneven, with a tattered appearance. This fringe decreases air turbulence and resulting noise. The feature is unlike most other birds whose wings have a sharp, clean edge. Second, the feathers covering the owl's wings, body, and legs are velvety soft. This serves to dampen and absorb the sounds of rustling air. Owls also have special curved wing feathers that allow flight at slow, quiet speeds.

A lessening of aircraft noise is of obvious benefit to the military, and stealth technology currently receives much attention. However, noise reduction is also important to commercial aviation. Major airports limit the amount of noise generated each day, for the benefit of nearby residents. The use of owl-feather technology may allow increased air

traffic without an increase in noise. Ideas include a retractable brush-like fringe on airplane wings, and a velvety coating on the landing gear. Studies show that these measures result in substantially less airplane noise. However, these additions also cause unwanted drag on the aircraft motion. Efforts continue to reduce both the noise and drag of aircraft by studying the highly successful owl.

Internet search words:
noise, owl, stealth

Questions for further study

1. What is the largest owl?
2. How many species of owl have been catalogued?
3. How is the loudness of sound measured?

A: pg. 200

Noise Reduction

Swift ↻→Aircraft Wings

T here are at least 70 species of swifts, among the fastest fliers in the bird world. They are graceful flyers, even sleeping and mating while on the wing. They may spend the night gliding a mile (1.6 km) above the ground. Aerial feeders, they catch insects on the wing and scoop water from ponds while in flight. The European swifts migrate round trip to Africa each year. Over a lifetime, swifts may fly a total of 2.7 million miles (4.5 million km), equal to 100 trips around the world, or six round trips to the moon. The impressive acrobatics of swifts are made possible by the adjustment or "morphing" of their wing shape. Dutch and Swedish scientists are studying the models of swift wings in wind tunnels. They find that extended wings work best for gliding, while

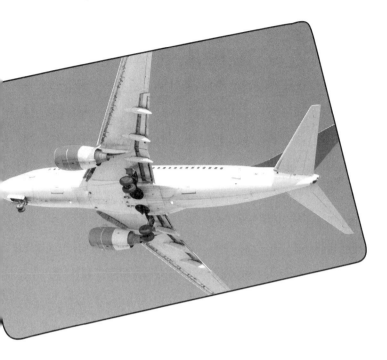

the swept-back, angled wings allow high-speed dives and quick turns. Wing morphing is of major interest to modern aviation. NASA is testing small aircraft with moveable wings for surveillance. From the Wright brothers until today, birds continue to teach us how to fly.

Reference

Lentink, D. and many others. 2007. How swifts control their glide performance with morphing wings. *Nature* 446(5733):1082–1085.

Questions for further study

1. How does evolution explain the origin of flight?
2. Since swifts do not perch on tree branches, where do they build their nests?
3. Are swifts and swallows the same?

A: pg. 200

Aircraft Wings

Toucan ⟳ Shock
Beak　　Absorber

T he toucan is a colorful bird in South America with an enormous beak. One species, named the toco toucan, *Ramphastos toco*, has a yellow orange bill reaching a length of nine inches (23 cm), a third of the bird's height.

The bird perches on sturdy portions of branches and relies on its long beak to reach tree fruits farther out. When active, it appears that the toucan might topple forward due to the awkward size of its beak. However, the beak is very light-weight while maintaining its strength, making up only 5 percent of the bird's total weight. The outer surface is made of keratin, the common protein material found in our own fingernails and hair. The keratin coating in the toucan beak consists of overlapping hexagonal layers that are somewhat flexible. This allows for bending and twisting motions of the beak. Meanwhile, the interior of the beak contains a foam-like, criss-crossed scaffold of tiny, flexible, lightweight bones. Some internal parts of the beak remain hollow, surrounded by the lattice of supporting bones.

The composite structure found in the toucan's beak provides a model with many possible uses. One potential application of such a structure is strong, lightweight safety helmets. Toucan beak "foam" also could provide protective cushion panels for cars and aircraft.

Reference

Eliot, John L. 2006. Power beak. *National Geographic* 210(6):30.

Questions for further study

1. In which countries might you find toucans?
2. Why are toucans so colorful?
3. How can the beak of a woodpecker survive hammering?

A: pg. 201

Shock Absorber

4

UNDERWATER LIFE

Of all the known varieties of life, about 60 percent live in fresh and marine waters. This is no surprise since our planet is 70 percent covered with water. The depths of the seas, averaging two miles deep (3.2 km) worldwide, remain largely unexplored. Explorers and pirates once sailed the oceans in search of treasure. Today, as we study underwater life, we find hidden treasures of useful design information.

Or speak

to the earth,

and it shall teach thee:

and the fishes of the sea

shall declare unto thee.

—Job 12:8

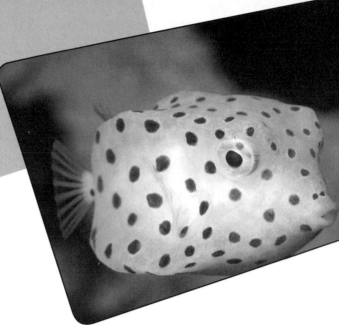

Boxfish ⟳Automobile Design

E ngineers from Mercedes-Benz and DaimlerChrysler look to nature for guidance in new automobile designs. One group of experts visited a marine aquarium to observe the streamlined efficiency of sharks. During their visit, however, they noticed a small, awkward-looking fish in the corner, a boxfish. The cube-shaped tropical fish slips through water with impressive efficiency. This is totally unexpected because, as the journal *Science* concludes, "One look at the aptly named boxfish, and you might expect it to swim as well as a barn would fly." Observations show that the boxfish swims easily and safely, even in turbulent water. Self-correcting vortices of current develop around its body and effectively cancel out the buffeting forces of surging water.

The skin of the boxfish consists of hexagonal, bony plates that give extra strength while minimizing weight. DaimlerChrysler has built prototype compact cars that copy the overall shape of the boxfish. The car's strong, lightweight door panels are patterned after the hexagonal

skin pattern of the fish. These test vehicles are found to excel in safety, comfort, maneuverability, and environmental friendliness. The compact cars also have outstanding fuel efficiency, measuring up to 70 miles per gallon (100 km/3.4 liters). The boxfish may look comical and awkward, but its design is a superb teaching tool for automotive engineers.

Questions for further study

1. How large is the boxfish?
2. Where is the boxfish found?
3. How many distinct species of boxfish have been catalogued?

Reference

Bartol, Ian. 2003. Boxy swimmers. *Science* 299(5608):817.

A: pg. 201

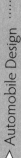

Echinoderms are a broad group of sea life including starfish and sea urchins. Members of the group are invertebrates that display a radial symmetry in their overall shape. Analysis of the bony structure

Brittlestar ⟳→Microlens

of a particular variety called featherstar, or brittlestar, shows remarkable optical properties. The animal's outer skeletal surface is covered with thousands of transparent microlenses, each about 0.0004 inches across. This size is ten times less than the thickness of a human hair. Each microlens focuses incoming light to nerve bundles located just beneath the surface of the skin. This compound eye covers much of the brittlestar body. The eye serves to detect nearby food, and it also shows a way of escape from predators.

Each tiny lens is composed of the clear mineral *calcite*. This crystalline mineral gives extra strength to the skeletal structure while providing vision in all directions. Calcite crystals ordinarily distort light rays that pass through, a blurring optical effect called birefringence. However, the brittlestar lenses are perfectly aligned to compensate and cancel this light distortion. Engineers at Bell Labs conclude that the design of the microlenses is advanced beyond any optical devices manufactured today. Further study of the optics of brittlestars may lead to improvements in the design of components for optical telecommunications networks.

In evolution theory, eyesight is assumed to have spontaneously developed on multiple occasions. This follows from the large variety of creatures that display vision. However, the detailed optics of the brittlestar, and every other visual creature, are advanced beyond our current understanding. The sense of sight is a clear example of planned, intelligent design in nature.

References

Aizenberg, Joanna, Alexei Tkachenko, Steve Weiner, Lia Addadi, and Gordon Hendler. 2001. Calcitic microlenses as part of the photoreceptor system of brittlestars. *Nature* 412(6849): 819–822.

Summers, Adam. 2004. How a star avoids the limelight. *Natural History* 113(4): 32–33.

Questions for further study

1. Where in the oceans are brittlestars found?
3. What does evolution theory suggest about the origin of vision?
2. Besides the lenses of brittlestars, where else might one find the mineral calcite?

A: pg. 202

Microlens

T he cuttlefish, a marine mollusk, is one of the most intelligent of all invertebrates. Cuttlefish range in size from two inches to three feet or more (5–90 cm). They live worldwide in tropical and temperate oceans. The swimming motion of cuttlefish results from waving or undulating continuous fins arranged along the sides of its body.

Cuttlefish ⟳ Camouflage

Of special interest, the cuttlefish displays the fastest color-changing ability of any known animal. Beneath its skin are many small elastic sacs, called *Chromatophores*, each sac filled with color pigments. Attached muscles expand or contract these sacs, changing the cuttlefish appearance in less than one second. The pigments include the colors brown, red, and orange. The cuttlefish is not limited to this color range, however. Deeper under its skin are white patches made of cells called *leucophores* that function as mirrors. These surfaces reflect the colors of the cuttlefish's nearby environment. When the cuttlefish swims beneath green seaweed, for example, it appears to instantly turn green. Military researchers seek to duplicate the cuttlefish camouflage ability. A rapid color-changing gel has been prepared, based on cuttlefish chemistry. This gel can be applied to military clothing and equipment.

Similar to the cuttlefish, squid likewise are masters of disguise by camouflage. In addition to their color changes, squid display an additional ability: they produce skin patterns that are polarized. This is a form of

light pattern noticeable only to eyes that are sensitive to polarized light. This is true of squid eyesight, but not for other mammals, including ourselves. The special ability suggests that squid may communicate detailed information to each other using polarized light patterns on their skin, even while they remain camouflaged and invisible to predators.

References

Hanlon, Roger. 2004. Beautiful and beastly squid. *National Geographic* 206(2):30–45.

Holloway, Marguerite. 2000. Cuttlefish Say It With Skin. *Natural History* 109(3):70–79.

Metz, Stephen, Editor. November 2006. Squid Hidden Messages. *The Science Teacher* 73(8):12.

Questions for further study

1. Name some other animals that change color.
2. Are cuttlefish used as seafood?
3. What is polarized light?

A: pg. 202

Camouflage

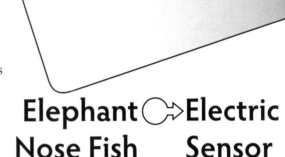

The elephant nose fish, 5–6 inches long (12–15 cm), is popular in fresh-water aquariums. It originates in the Congo River of West Central Africa. This unusual fish is able to generate and detect weak electric fields. Such fields result from nearby positive or negative electrical charges.

Elephant Nose Fish ⟳→ Electric Sensor

The elephant nose fish quickly notices prey, predators, and potential mates in its vicinity by the small changes they cause in the surrounding electric field, also called the *aura*. The fish has a highly efficient, complex sensing apparatus that measures the tiny nearby disturbances.

One practical application learned from the elephant nose fish involves automobile air bags. These devices save many lives during crashes, but they can prove dangerous to small children. There is a need for air bag deployment to be sensitive to the size of the seat occupant, whether child or adult. A surrounding electric field provides an ideal solution. A weak, harmless electric field can be generated by electronics when the car is operating. A passenger disturbs this electric field, depending on body size. In a stopping emergency, when a child is present, the airbag deployment can automatically be lessened or deactivated.

There are further applications learned from the elephant nose fish. For example, electric fields are useful for the robotic sensing of objects, and also for the detection of intruders in homes. Eventually, our electric field technology may approach that which the elephant nose fish displays.

Another popular aquarium fish is called the brown ghost knifefish. About six inches long, it also produces electric signals to detect obstacles and to communicate with other knifefish. This amazing creature is also able to "jam" or distort the signals of rivals. It generates an electric pulse with a frequency that mimics that of nearby competitors. This pulse confuses the sensing ability of other knifefish, which then swim away. Military submarines and airplanes use similar electronics jamming strategy devices for their defense.

References

Milius, Susan. 2005. Tszzzzt! *Science News* 168(21):324.
Weiss, Peter. 2004. Sixth sense. *Science News* 165(25):392–393.

Questions for further study

1. What is an electric field?
2. Does an electric eel also generate an electric field?
3. What large sea animal has a nose protrusion somewhat similar to the elephant nose fish?

A: pg. 203

Electric Sensor

Ship designers study the movement of fish to improve the propulsion of sea vessels. One important goal for ships is a smooth, streamlined path through the water using minimal energy. It is noticed that fish have two

Fish ⟳ Ship
Motion Propulsion

distinct advantages for their propulsion. First is a skin coating of mucus that cuts down on drag and frictional energy losses. Researchers are experimenting with similar slippery surface materials for boat hulls. One promising chemical, polyethylene oxide, has long molecular threads that suppress the turbulent eddy currents of nearby moving water. Continued studies of the composition of fish mucus may help us arrive at improved artificial coatings for boats and ships.

A second major fish advantage for propulsion is the fin and tail structures that are much more efficient for thrust than the common screw propeller or paddlewheel. Creatures such as penguins and humpback whales also display excellent speed and maneuverability in water by using their flexible flippers. Efforts are underway to develop fin-propelled watercraft. The challenge is to discover the best fin shape, and also to find materials with a permanent elasticity approaching that of fish tissue. Wind tunnel studies have been applied to the serrated or scalloped edges of whale flippers. It is found that lift is increased and drag is diminished when compared with smooth edges on flippers. This unexpected

discovery also has potential application for improved airplane propellers and helicopter rotors.

Competitive swimmers benefit from the study of fish design. Olympic swimsuits typically have a fabric surface that mimics the skin of sharks. This fabric has tiny v-shaped ridges, called *dermal denticles,* that reduce water turbulence and drag. Sea creatures reveal to us the secret of swimming success in their water world.

References

Ashlet, Steven. 2004. Bumpy flying. *Scientific America* 291(2):18–19.

Frazier, Kendrick, Editor. 1976. Fins and mucus for boats? *Science News* 110(20):314.

Lock, Carrie. 2004. Ocean envy. *Science News* 166(10):154–156.

Questions for further study

1. How does the speed of Olympic swimmers compare with fish?
2. Is it known why whales breach, or leap above the water?
3. What is the supposed evolutionary origin of whales?

A: pg. 203

Lobster ⟳→ Telescope
Eye   Lens

Lobsters appear awkward with their protruding eyes, antennae, and large claws. However, these creatures have much to teach us, including the unique design of their eyes. Most animals — and people as well — focus light by refracting or bending incoming light rays through the eye's cornea and lens. In contrast, the lobster eye works by the reflection of light from tiny, flat mirror-like surfaces. Its eye consists of thousands of rectangular tubes arranged on the outer eye surface. Light enters these small openings and reflects inward off the shiny inside surfaces. Precise alignment of the mirrors directs the separate light rays so that they focus together on the retina receptors. The eyes of shrimp and prawns are somewhat similar to those of lobsters.

The lobster's eye design has been copied in a new generation of x-ray space telescopes, sometimes called "lobster-eye instruments." X-rays are a very energetic form of radiation emitted by stars. Ordinary lenses or

Lobster Eye

⤳ Telescope Lens

mirrors are not suitable for the focusing of x-rays. The problem is that x-ray radiation passes directly through ordinary mirrors, unless it hits at a small glancing angle, whereupon it reflects. This grazing angle is exactly how light enters the lobster-eye telescope lens. For laboratory use, the lobster lens can also function in reverse fashion to generate an outgoing, parallel beam of x-rays. The advanced optical design of the lowly lobster is beneficial to x-ray research in space.

Reference

Chown, Marcus. 1996. I spy with my lobster eye. *New Scientist* 150(2025):20.

Questions for further study

1. What is an x-ray?
2. What do x-ray telescopes see?
3. Is the crayfish eye similar to that of a lobster?

Telescope Lens

A: pg. 204

Mussels ↻ Adhesive

Mussels are marine animals that firmly attach themselves to rocks, shells, piers, ship hulls, or each other. The adhesive produced by mussels is remarkably strong and durable. It functions in turbulent salt water, and like all other materials from nature, it is biodegradable. Laboratory efforts to duplicate this mussel glue have been only partially successful. One surprising mussel-glue component is the element iron. Such metal atoms have not previously been found in similar biological functions.

A mussel stretches dozens of tiny filaments from itself to a surface, attaching each strand with a dab of glue. The *polyphenolic* protein, a key glue ingredient isolated from mussels, has been used to bond human tissue after surgery. This technique has advantages over the traditional use of sutures. Further understanding of the mussel's bio-adhesive glue may

lead to precise surgical sealants in eye surgery. A biotech company named *Nerites*, located in Wisconsin, is developing *Medhesive* for this purpose. Also, a chemical understanding of the mussel glue could lead to solvents that will dissolve it when needed. Applications may include antifouling paints to discourage underwater adhesion to surfaces by barnacles.

References

Benyus, Janine. 1997. *Biomimicry: Innovation Inspired by Nature*. Harper Collins Publishers, New York.

Mitchinson, Andrew. 2006. Mussel muscle. *Nature* 442(71050):877.

Questions for further study

1. Why are mussels a problem in inland waters?
2. What does the term *biodegradable* mean?
3. Who invented superglue?

A: pg. 204

Adhesive

Octopus ⟳ Robotics

The octopus is a very skillful animal. With eight long, flexible tentacles it can gently wrap around nearby objects. However, the creature has been observed to stiffen an arm when necessary to transfer prey from one location to another. The arm becomes articulated into short, straight sections that are quasi-jointed, similar to the human arm. Scientists recognize that the octopus arm may illustrate the optimum solution for point-to-point movement of robotic arms. In space, for example, a stiff arm tends to push objects farther away rather than

drawing them inward. In contrast, a flexible robotic arm could wrap around objects to move or retrieve them. There remains the technological challenge of building an artificial arm that mimics the ability of the octopus. See the "muscles" discussion in chapter 6 for a possible solution.

Reference

Sumbre, G., G.Fiorito, T. Flash, and B. Hochner. 2005. Neurobiology: motor control of flexible octopus arms. *Nature* 433(7026):595–596.

Questions for further study

1. Does an octopus regenerate a lost arm?
2. How long can octopus arms grow?
3. What kinds of robotic arms are currently used in space?

A: pg. 205

Robotics

Many seashells are made of composite, or multiple, materials. They have a layered internal structure that results in hardness, strength, and flexibility. Perhaps you have noticed the thin appearance

Seashell ⟳ Construction Material

of many shells. Yet they easily survive the turbulence of underwater life. The shell composition, also found in many bones, is described as a "bricks and mortar" structure. The "bricks" are tiny crystals of calcium carbonate, $CaCO_3$, held in place by the "mortar," a network of proteins. The combined materials hold firmly together to comprise the thin shell with about a million microscopic layers. The resulting shell is called *nacre* or *mother-of-pearl*. The "hard" bricks of calcite and the "soft" protein mortar are complementary in their response to stresses and strains. They are stronger when bonded together than either one separately.

Materials scientists have copied the composite shell structure with limited success. One method is to dip a glass slide alternately into a clay solution (the bricks) and then a cementing polymer (the mortar). After a few thousand such dips, a strong composite layer builds up. Alumina, Al_2O_3, is also sometimes used for the "brick" portion of the layers. An alternate fabrication approach involves the repeated freezing of a mixture of water and ceramic dust particles on a flat surface. This results in multiple thin layers of ice and ceramic. Later, the ice is removed by freeze-drying and replaced with glue. Some of the manufactured

Seashell

structures are centimeters thick when completed. Eventual medical applications include the fabrication of biological hard tissue and artificial bone. On a larger scale, scientists are hopeful of improved body armor and the manufacture of extremely strong composite components for aircraft and automobiles.

References

Gorman, Jessica. 2003. Material mimics mother-of-pearl in form and substance. *Science News* 163(25):397.

Rubner, Michael. 2003. Synthetic sea shell. *Nature* 423(6943):925–926.

Summers, Adam. June 2006. Tough as shells: a promising candidate for artificial bone. *Natural History* 115(5):28–29.

Weiss, Peter. 2006. Mother-of-pearl on ice. *Science News* 169(4):51–52.

Questions for further study

1. What is the origin of the name *mother-of-pearl*?
2. What materials are currently used for artificial bone?
3. What are some traditional uses of shell material?

A: pg. 205

Construction Material

Sea Slug ⟳→Chemicals

Sea slugs are colorful snail-like marine animals with fringe-like projections instead of shells. They range from an inch to two feet in size (1–60 cm). When threatened by predators, some sea slugs eject protective, inky chemicals into the water. Researchers find that the *Aplysia* sea slug stores component chemicals separately in its glands. When disturbed, the rapid mixing of these components forms hydrogen peroxide, ammonia, several acids, and purple dye. This unpleasant mixture is sprayed outward, discouraging predators from dining on the sea slug.

An additional chemical protein has been detected in sea slugs that is anti-microbial in nature. Scientists believe this material repels certain predators, and it also functions as a salve for wounds that the sea slug

may receive. There is hope that the antibacterial protein may be useful for our own healthcare. The sea slug protein may also serve to prevent the growth of bacteria and other unwanted microbes in community water supplies.

Internet search words: sea slug, antimicrobial

Questions for further study

1. How many species of sea slug are known to exist?
2. What does the term *microbe* refer to?
3. What are some descriptive sea slug names?

A: pg. 206

Sea Slug

Chemicals

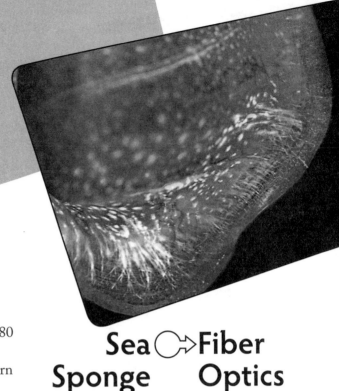

T he Venus
Flower
Basket is a sea
sponge that
typically lives 3,280
feet (1,000 m)
deep in the western
Pacific Ocean.
They also tolerate
lower water pressure and

Sea ⟳ Fiber
Sponge   Optics

are sometimes displayed in saltwater aquariums. The scientific name
is *Euplectella aspergillum*. This sponge grows its own optical fibers,
as thin as human hair, then lights them up like an undersea lamp to
attract nearby prey. The optical fibers have at least two advantages over
man-made fiber optics strands. First, the sponge fibers grow at the
temperature of ocean water, whereas commercial fibers are produced in
high-temperature, glass-melting furnaces. As a second advantage, the
sponge fibers are coated with an organic sheath that toughens them. As
a result, they can be twisted and even tied in a knot without breaking.
In contrast, a major problem with commercial optical fiber is its fragile
nature. Scientists at Bell Labs are studying the Venus Flower Basket for
clues in the manufacturing of strong commercial fiber optics cables.

In traditional Asian cultures, a dried Venus Flower Basket is a
symbolic wedding gift. When alive, the living sponge typically houses
two small shrimp, male and female, who spend their lives held captive

inside the hollow mesh tube. Their offspring escape the tube and find a new sponge home of their own. The enclosed shrimp keep the Flower Basket clean. In return, the glowing strands of fiber attract small organisms which drift inside the tube as food for the shrimp. This behavior is called *symbiosis*, whereby distinct organisms benefit from mutual cooperation.

Reference

Benyus, Janine. 1997. *Biomimicry: Innovation by Nature*. Harper-Collins Publishers, New York.

Questions for further study

1. What is the composition of man-made optical fiber?
2. Have fossil sponges been found?
3. Besides the mutual benefit between the Flower Basket and shrimp, give some other examples of symbiosis.

A: pg. 206

Fiber Optics

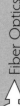

Whale ⟳ Submarine

Designers of submarines face many challenges. These include cold, darkness, and extreme water pressure at great depths. In addition, communication and location of one's position become difficult when the vessel is submerged. In contrast, whales cruised the depths of the seas long before the first submarines were launched. These magnificent creatures display successful solutions to the problems of submarine travel.

It is no coincidence that submarine shape resembles that of whales. Consider the long, streamlined hull of a submarine, and also its fins, or hydroplanes, that provide stability and steering ability. Sperm whales have a blunt nose that looks awkward at first sight. However, nuclear submarines have a closely similar bow. Both cruise through water at similar speeds. This may average 12 knots (14 mi/hr or 22 km/hr), although both whales and submarines can travel much faster when necessary. The blunt nose results in streamlined, efficient movement through water with minimal noise or water turbulence.

Submarines also use rear propulsion with propellers, somewhat equivalent to the whale's thrusting tail. Even the whale's sonar system (sound navigation and ranging) is utilized by modern submarines.

Whales have been present since day 5 of the creation week when flying and swimming creatures appeared. They are truly majestic witnesses of creation.

Internet search words: whale, submarine, design

Questions for further study

1. What is the largest whale?
2. How deep can whales dive?
3. How do whales communicate with each other?

A: pg. 206

Submarine

Whale

5

LAND ANIMALS

This chapter describes several creatures that may be more familiar than other book entries like Namib beetles and sea slugs. Examples include deer and tree frogs; both make important contributions to biomimicry. In fact, every item on the earth, both living and nonliving, provides us with practical object lessons. The following creatures range from dinosaurs to deer. Even a companion dog offers lessons in useful design.

And out of the ground the LORD God formed every beast of the field, and every fowl of the air; and brought them unto Adam to see what he would call them: and whatsoever Adam called every living creature, that was the name thereof .

— Genesis 2:19

Ankylosaurus ⟳
Fiberglass

There is a common perception that animals living in the distant past were primitive and simple. However, fossil studies show just the opposite. Consider the ankylosaurus dinosaur. This creature grew to 33 feet (10 m) in length and displayed impressive bony plates along its back. It also carried a club-like tail for defense. However, this armor plating was unlike the solid bony structure of today's crocodiles. Instead, fossil evidence reveals that the dinosaur's plates and protrusions consisted of multiple layers of collagen fibers. The fibers crisscrossed each other at right angles, an arrangement similar to that of modern fiberglass. Fiberglass is often used when great strength is needed, while limiting the weight, as in boat hulls or surfboards. The bulletproof vest material Kevlar has similar interwoven fibers that safely absorb impact forces and energy.

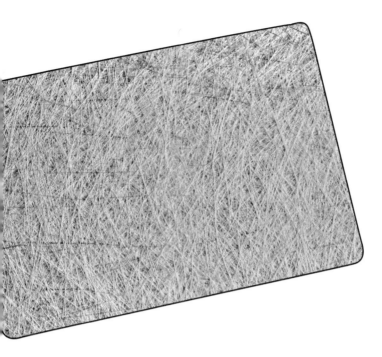

The ankylosaurus was endowed with its own form of organic fiberglass. The result was great strength and protection from predators. Further research may lead to improvements in strong, lightweight materials. One proposal is the construction of strong, protective barriers along highways. The ankylosaurus may be gone, however, its fossils reveal useful information.

Internet search words:
  ankylosaurus, armor,
  fiberglass

Questions for further study

1. How large were dinosaur eggs?
2. How many dinosaur species have been discovered?
3. When was fiberglass invented?

A: pg. 207

Fiberglass

Antler ⟳ Organ Repair

R eptiles and amphibians have the ability to re-grow certain body parts. When a lizard loses its tail, for example, a new tail grows back. People, and mammals in general, are not able to perform this kind of body repair. If a person loses a finger in an accident, for example, the loss is permanent unless the finger can be reattached. However, there is one nearly unique example of re-growth in mammals. This refers to the antlers of deer and related animals. Antlers are a solid bone material that sheds annually and then grows back in the spring. The rapid growth rate can measure nearly an inch (2.5 cm) per day. In contrast, our fingernail growth is one to two inches a year, hundreds of times slower.

Scientists are uncovering the internal chemical signals that trigger the annual regeneration of antlers. Stem cells appear to be directly involved in this process. There is a striking similarity between antlers and

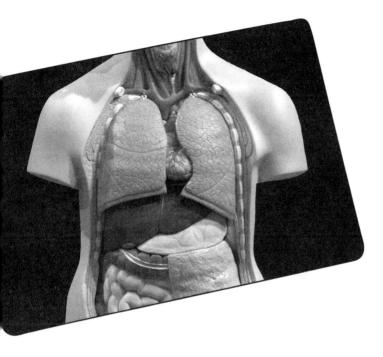

human limbs. Both consist of bone, cartilage, skin, blood vessels, and nerves. Further understanding of antler growth may lead to the repair of human tissue and organs. There may also be new treatments for arthritis, osteoporosis, and immune disorders.

Internet search words:
antler, regeneration

Questions for further study

1. Which animal holds the record size for antlers?
2. How is antler size related to animal age?
3. What are stem cells?

A: pg. 207

Organ Repair

Dog Paw ⟳➔ Shoe Soles

In 1935, inventor Paul Sperry sought a solution to a problem encountered in his hobby of sailing off the shore of New England. Whenever the boat deck became wet, it was slippery and dangerous. One winter day during a walk, he noticed that his cocker spaniel remained surefooted, even on slippery sidewalks. Sperry later examined the dog's paws closely and noticed wave-like grooves on the pads.

Sperry began to experiment with shoes. He obtained a thick sheet of rubber and cut grooves in a zigzag "herringbone" pattern. Then he attached two sections of this rubber to his canvas, flat-bottomed sailing sneakers and tested them. The traction was obvious when he walked on ice or any slippery surface. The grooves allowed the shoe sole to deform

slightly and to grasp the ground surface. When the surface was wet, the grooves channeled water outward from under the shoe. Sperry went on to manufacture the first non-skid deck shoes, called *Sperry Top-Siders*. They were an immediate success in the world of sailing and are still manufactured today in a style called the *Authentic Original*. Grooved soles on sports shoes are now a worldwide standard.

Internet search words:
    dog's paw, Paul Sperry

Questions for further study

1. Where else besides shoe soles is the herringbone pattern found?
2. How have Paul Sperry's ideas been applied to automobiles?
3. Where can one learn more about Paul Sperry's discovery?

A: pg. 208

Shoe Soles

Gecko ⟲⟶Adhesive

Gecko is the common name for a large variety of tropical lizards. These acrobats have the ability to run upside-down across ceilings while hunting for insects. How are geckos able to overcome gravity and cling to smooth surfaces? A microscope shows that their toes are equipped with a carpet of scales holding a half-million tiny brush-like projections called *setae*. Each of these seta then branch further into finer hairs. The result is a weak attractive force between each "split end" and any surface, no matter how smooth. Two "sticking" forces, called van der Waals bonds and capillary attraction, dominate on this micro scale. Geckos utilize these attractive forces to hold firmly to almost any surface. In walking, the lizards easily detach their feet by bending or peeling them away from the surface, like post-it notes. In experiments, tiny backpack weights are attached to the geckos to test their holding ability. Measurements show that geckos stick firmly on the ceiling, even when extra weight is applied, as much as 400 times greater than the lizard's own weight. The lizard's method of adhesion is clean, reliable, and works equally well on wet or dry surfaces.

Scientists are learning to duplicate the lizard's ability with "gecko tape." A person wearing a glove covered with this tape could actually dangle from the ceiling. The tape consists of microscopic plastic hairs mounted on a flexible base. In tests thus far, the plastic adhesive loses its holding ability after only about five contacts because stray molecules

Gecko

Adhesive

of water on the surface or in the air cause the artificial bristles to clump together and become ineffective. The gecko lizard solves this moisture problem by incorporating water-repelling keratin within its hair follicles. Keratin is a common protein material found in feathers, hair, and fingernails. As gecko tape is further improved, it may be useful in surgery to bond tissues together. Automobile tires with greatly increased traction may also result, along with glue-free tape, climbing gear, and building materials. And just for fun and exercise, imagine walking across the ceiling on your hands!

References

Graham-Rowe, D. 2003. Fancy a walk on the ceiling? *New Scientist* 178(2395):15.

McDonagh, Sorcha. 2003. Caught on tape. *Science News* 163(23):356.

Questions for further study

1. Explain the nature of van der Waals and capillary forces.
2. Who invented post-it notes?
3. How do spiders and other insects walk on ceilings?

A: pg. 208

Adhesive

I f you quickly stand from a prone position, you may notice your heart's effort to send blood to your elevated brain.

Giraffe ↻→Antigravity Spacesuit

For a moment you may see twinkling stars or feel faint. This feeling is magnified for astronauts who spend extended time in space, where the gravity force is near zero. As a result, the heart no longer needs to pump blood "uphill" to the brain against gravity. Instead, blood may pool around the heart and thorax, a condition known as fluid shift. Back on earth, the body slowly returns to its normal functions in response to gravity. However, some astronauts are plagued with fainting spells for several days. The return of the gravity force causes blood to pool in the lower extremities, limiting circulation to the astronaut's brain.

NASA scientists are solving the blood pressure problem by studying the giraffe. As a giraffe raises and lowers its head by several feet, large blood pressure changes might be expected in its brain. However, giraffes do not experience fainting problems. Instead, a series of circulation valves in the neck prevents major blood pressure changes in the giraffe's head. Also, the giraffe's legs have especially tight skin and strong muscles. These features prevent blood from pooling in the long legs of the giraffe.

Astronaut spacesuits and military flight suits control blood circulation using designs based on the giraffe. For example, when experiencing acceleration, pressure is temporarily increased around a person's legs to prevent blood buildup. In effect, the blood is pushed upward toward the brain to prevent fainting. The giraffe shows us how to keep pilots and astronauts safe.

Reference

Reebs, Stephan. October 2006. Uphill Battle. *Natural History* 115(8):14.

Questions for further study

1. What exercise activity of orbiting astronauts counters "fluid shift"?
2. How tall is an adult giraffe?
3. How large is the giraffe heart?

Antigravity Spacesuit

A: pg. 208

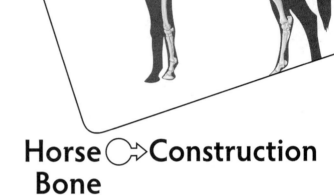

Horse ⟳ Construction Bone

The lower bone in a horse's leg, called the third metacarpus, is the thickness of an adult person's wrist. One side of this bone has a pea-sized hole, called the foramen, where blood vessels enter the interior of the bone. This is surprising since similar holes in engineered structures are a frequent reason for material weakness and failure. It appears at first sight that the leg of the horse shows poor design. However, this would be the opposite conclusion from the many positive design examples in this book. Scientists have taken a close look at the horse bone, and they find very beneficial features.

The foramen opening is surrounded by flexible material that directs stresses toward stronger regions of the leg bone. Occasionally a horse experiences a broken metacarpus, a special hazard in racing. When this occurs, however, the fracture usually does not take place at the site of the foramen opening, but elsewhere along the bone. Engineers have modeled the horse bone by drilling holes in plates of various materials. They then surround the openings with polyurethane foam and apply force. Under extreme stress, the openings are *not* the source of breakage or failure, similar to the experience of horses. The foramen definitely is not a weak part of the horse metacarpus bone.

Traditionally, openings or holes in manufactured structures are reinforced with an extra thickness of material. Think, for example, of a porthole on a ship with its surrounding ring-shaped plate and bolts. The design of the horse metacarpus suggests an alternative approach of varying the type of material used around openings in structures. The result is additional strength while decreasing weight.

Internet search words:
design, foramen, horse

Questions for further study

1. Do we have any bones with a foramen opening, similar to the horse?
2. What is the top speed of a race horse?
3. How much stress is generated in the bones of athletes?   A: pg. 209

Construction

Penguin ○⇢ Sunglasses Eye

Penguins have clear vision in spite of the intense glare of polar sunlight. These Antarctic animals are found to have an external eye fluid that filters blue and ultraviolet colors from the solar spectrum. The result is clear vision while protecting the eye from injury. Birds of prey, including eagles, falcons, and hawks, also produce the effective retinal fluid found in the optic systems of penguins.

Research has duplicated the optical advantage of penguins with an orange-colored dye or filter, leading to several applications. Many welders now use orange-colored masks or screens that are more transparent

and safer than the old-style dark masks that obstructed vision. Orange-tinted sunglasses give pilots, sailors, and skiers improved vision in bright sunlight, haze, or fog. There is also the promise that orange-tinted glasses may help patients suffering from visual loss due to cataracts or macular degeneration.

Internet search:
  Suntiger.com

Questions for further study

1. Where in the world are wild penguins found?
2. How large are penguins?
3. What is it about the color orange that makes a useful light filter?

A: pg. 209

Sunglasses

Tree ⟳ Automobile
Frog   Tires

Tree frogs are able to cling to the undersides of slippery, wet leaves. Yet their holding ability is distinct from that of the gecko, discussed earlier in this chapter. While the gecko has many dry hair follicles on its feet, frog feet have microscopic bumps raised above a thin mucus film. These bumps or "cleats" make contact with the leaf surface and cling by what is called "dry friction." This attraction involves molecular forces similar to those utilized by the gecko. In addition, the mucus sticks to the leaves by "wet adhesion," an additional attractive force. This is also how wet paper sticks to a glass surface such as a window. When a frog is on a wet leaf, excess water is carried away by channels between the cleats on its feet. The dual mechanisms of cleats

and mucus allow tree frogs to hang upside-down on leaves, regardless of whether the surface is dry, wet, smooth, or rough. Researchers believe that tree frog studies will lead to improved automobile tire designs that wick away water while maintaining traction.

Reference

Jaffe, E. 2006. Walking on water. *Science News* 169(23):356.

Questions for further study

1. In what regions do tree frogs live?
2. How do tree frogs act as barometers?
3. How many species of frogs are known in the world?

A: pg. 210

6

PEOPLE

The human body has its share of infirmities, yet it is a marvel of intelligent design and planning. Two centuries ago, British scholar William Paley (1743–1805) wrote about the intricacies of human vision. He concluded that just as a watch requires a watchmaker, a designed eye requires an eye maker. Paley was correct! In recent years our brain has been compared to an advanced computer, far surpassing any present or planned computer system. We are still learning much about our fearful and wonderful creation (Ps. 139:14), as illustrated by the following design details.

> The marks of design are too strong to be got over. Design must have had a designer. That designer must have been a person. That person is GOD.
>
> — William Paley

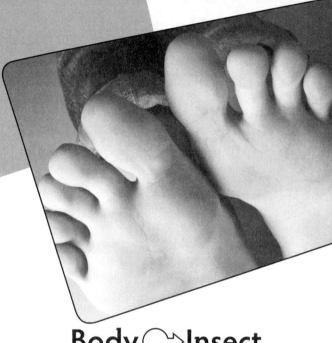

Body ⟳ Insect
Odor Repellant

Some people attract mosquitoes like magnets while others are rarely bothered or bitten. These pesky insects zero in on a combination of carbon dioxide and body odorants. We know that mosquitoes also are drawn to exhaled carbon dioxide. However, British research shows that some people give off "masking" odors that prevent mosquitoes from locating them. Still other individuals produce natural repellants that mosquitoes avoid. As a test, mosquitoes were placed in the base of a Y-shaped tube. The insects made a choice of direction when moving into the upper branches of the tube. A gentle air flow down one side was clean, while the other side was laced with odor from the hands of volunteers. The mosquitoes honed in on the odors of some volunteers, but not others. The mosquito-free volunteers had somehow switched

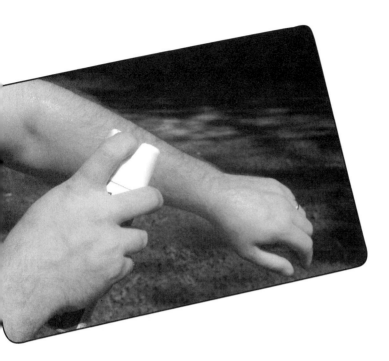

off the attraction of mosquitoes. The researchers found at least eleven distinct repellant compounds produced by the volunteers. The laboratory identification of these chemicals may lead to effective, natural insect repellants. Such natural chemicals could bring insect relief to people, pets, and livestock.

Internet search words:
  body odor, mosquitoes,
  repellant (repellent)

Questions for further study

1. Which mosquitoes bite, the males or females?
2. What chemicals are typically used in insect repellants?
3. What is a major worldwide danger of mosquitoes?

A: pg. 210

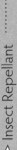

Insect Repellant

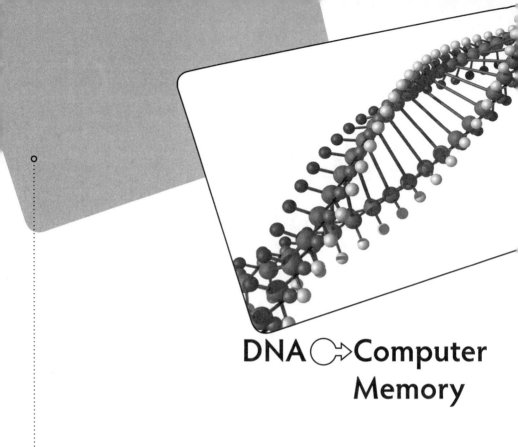

DNA⟳Computer Memory

DNA is a complex biomolecule that is embedded in the cells of all living organisms. DNA consists of two intertwined chains of connected atomic molecular structures. The arrangement of smaller units spaced along the DNA chair serves as a blueprint or recipe for the composition and internal activity of plants, animals, and people. Life is incredibly complex, and detailed information is needed for its functioning. Just a single gram of DNA holds as much information as one trillion compact discs. Obviously, such computer storage capacity lies far beyond current technology.

Computer experts look to artificial DNA for future computer memory needs. A team in Japan has experimented with the DNA of living bacteria. They placed binary information into segments of the bacterial DNA. The information includes numbers and alphabetical

letters. At a later time, a process called gene sequencing is able to retrieve this information. One of the first items of stored information was the simple message "E = mc² 1905." This refers to Albert Einstein's famous equation, published in 1905. The future of computer memory may well involve "designer" versions of the biomolecule DNA, already present in all of our cells.

Reference

Yozomu, Yachie, et al. 2007. Alignment-based approach for durable data storage into living organisms. *Biotechnology Progress* 23:501–05.

Questions for further study

1. What does DNA stand for?
2. What is the size of a DNA molecule?
3. What do the letters stand for in Einstein's formula $E = mc^2$?

A: pg. 210

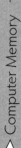

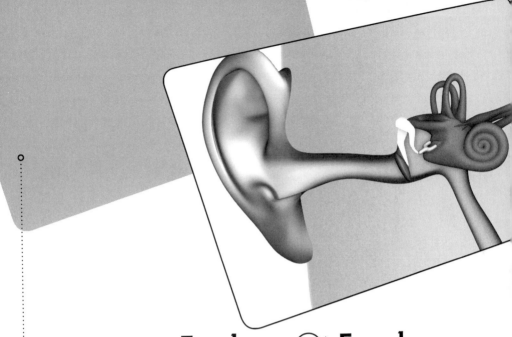

Eardrum ⟳ Earphone

The eardrum is a marvel of engineering. As thin as tissue paper, it vibrates in response to the slightest changes in air pressure. If the eardrum surface moves inward a distance equal to the diameter of a single atom, 100 millionth of a centimeter, a distinct sound is perceived. Clearly, a healthy eardrum is ultra sensitive.

Alexander Graham Bell (1847–1922) was a professor of vocal physiology at Boston University. At this time, electronic communication was limited to the dots and dashes of Morse Code. In his research, Bell looked for ways to transmit the various frequencies or vibrations of the human voice. His study was motivated by the deafness of his mother and also his wife. Bell soon realized that the human ear provided the key to success. Bell noted that the bones of the inner ear amplified the slight vibrations of the eardrum. The signal then passed into the liquid-filled cochlea, or inner ear, where electrical signals were directed onward to the brain. In Bell's words, "It occurred to me that if a membrane as

thin as tissue paper could control the vibrations of bones that were, compared to it, of immense size and weight, why should not a larger and thicker membrane be able to vibrate a piece of iron in front of an electromagnet...?" Bell was describing the principle of both the microphone and the loudspeaker. From earphones to cell phones, voice and music transmission is modeled after the design of our eardrum.

Internet search words:
Alexander Graham
Bell, ear, telephone

Questions for further study

1. What were some of Alexander Graham Bell's other inventions?
2. What are the names of the three bones in our inner ear?
3. How was Alexander Graham Bell honored at his death on August 14, 1922?

A: pg. 211

Earphone

Eye Iris ⟳→Identification

F ingerprints have long provided unique personal identification. The prints have more than 35 measurable characteristics that can appear in almost limitless combinations. There is one other personal feature, however, with far greater potential for unique recognition. This is the iris of our eye, the blue-green-brown component that controls the amount of entering light. Look closely in a mirror and your iris will show a large number of star-like points. A single iris has at least 266 identifiable characteristics, perhaps the most data-rich physical structure on our bodies. While fingerprints can be hidden or altered by intentional scarring, the iris is beyond tampering. There is an estimated one chance

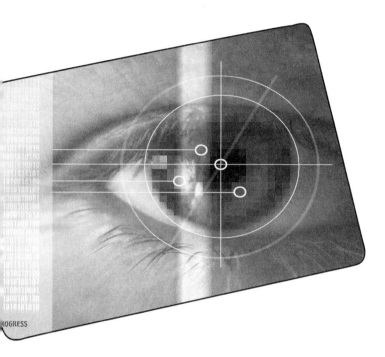

in 10^{78} that the irises of two people will exactly match. The earth's population is about 68 zeros smaller, seven billion people. In other words, the probability of two identical irises is negligible. Scanning of the iris may someday be commonplace for secure identification purposes.

References

Atick, Joseph. 2004. Face of the future. *Kiplingers* 58(12):26–27.

Stone, Brad. November 30, 1998. Tired of all those passwords? *Newsweek* p. 12.

Questions for further study

1. How does the iris control incoming light to the eye?
2. What are some ways to express the number 10^{78}?
3. Where is iris recognition currently used for security?

A: pg. 211

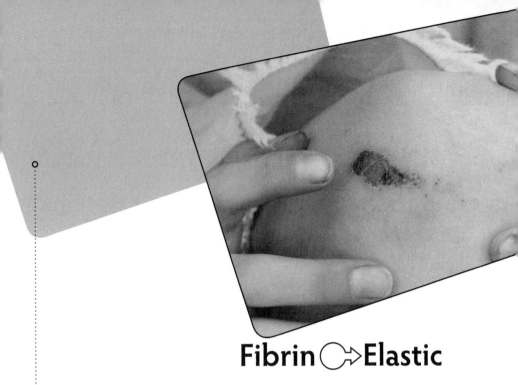

Fibrin ↻→ Elastic

Nearby vessels to a wound are sealed off with a clotting mechanism to stem the flow of blood. This happens in spite of the blood pressure generated by our beating hearts. A protein called fibrin is the key to this essential process. A clot results when the fibrin forms a sticky web embedded with fragments of body cells. Researchers have explored the flexibility of the fibrin fibers. Individual fibers were grasped with a device, and a tension force was applied. The fibers were found to

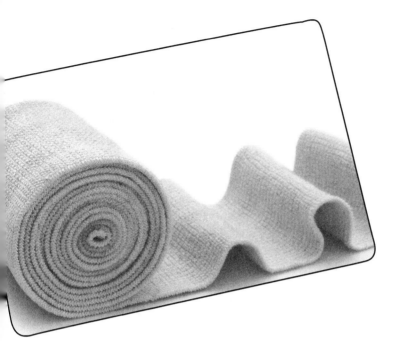

stretch to more than four times their initial length, and then spring back to original size. No other protein fiber in nature is known to have this extreme stretching ability, including spider silk. The chemical fibrin serves as a useful model for making strong elastic materials.

Reference

Liu, W. and many others. 2006. Fibrin fibers have extraordinary extensibility and elasticity. *Science* 313(5787):634.

Questions for further study

1. What is hemophilia?
2. How does Coumadin affect blood clotting?
3. How is the blood clotting mechanism related to intelligent design?

A: pg. 212

Elastic

Fingerprint ⟳→Prosthetic Hand

Our fingerprints are useful to law enforcement, but the intricate grooves and swirls have long been a mystery to scientists. Certainly the tiny ridges help our grip on slippery objects, but there is much more. Studies in a Paris laboratory now give another purpose: fingerprints greatly improve our sense of touch. A grooved surface was prepared, similar to our fingertips, and then rubbed across various textured surfaces. Rapid micro-vibrations resulting from the artificial grooves were electronically recorded. The fingerprint was found to amplify vibrations in the exact frequency range where our sense of touch is most sensitive. In fact, a grooved surface produced 100 times more sensitivity than a similar smooth surface.

Texture information from our fingertips helps us identify objects by touch. Even the elliptical swirls of fingerprints serve a purpose: they allow helpful vibrations to stimulate our nerves, no matter which way the finger moves across an object. The fingerprint studies may lead to improved artificial hands. Similar grooves imprinted on prosthetic fingers could produce vital electronic tactile feedback for the wearer.

References

Miller, Greg. 2009. Fingerprints enhance the sense of touch. *Science* 323:572–573.
Sanders, Laura. 2009. Fingerprints filter vibrations. *Science News* 175(5):10.

Questions for further study

1. Is every fingerprint unique?
2. What typical frequencies do our fingerprints detect?
3. Do primates have fingerprints similar to people?

A: pg. 212

Prosthetic Hand

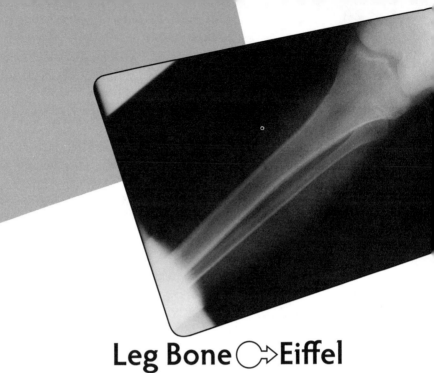

Leg Bone ⟳→Eiffel Tower

T he Eiffel Tower has stood above Paris for over a century. Including the antennas, the tower reaches a height of 1,063 feet (324 m). The base of this elegant wrought-iron structure follows the design of the human femur or thighbone. The upper ball joint fits into the hip socket, and is unusual in that it is located off-center from the bone. In the 1850s, anatomist Hermann von Meyer studied the lattice of bone ridges within the femur head, called *trabeculae*. Swiss engineer Karl Cullman later generated a mathematical model of the femur design. He found that the trabeculae functioned as a series of curved studs and braces. Resembling fibers, the strands of bone cross each other in layers at right angles. They are precisely lined up to carry forces of tension and compression. As a result, the fibers efficiently support and transfer the off-center weight of the person.

A French engineer named Gustave (Gustavo) Eiffel used a model of the femur model to design the tower that bears his name. There are large outward flares at the base of the Paris landmark. A lattice of steel studs

and braces within the curves supports the off-center weight above. This successful architectural design at the base of the Eiffel Tower duplicates the upper end of our femur bones. The tower was originally built as a temporary structure for a Paris Exposition or World's Fair in 1889. Competing architects scoffed at the tower and predicted it would soon collapse under its own weight. It remains today, over a century later, the destination of millions of tourists annually.

Internet search words:
Eiffel Tower, femur,
Herman von Meyer,
Karl Cullman

Questions for further study

1. What is the largest bone in the human body?
2. How much does the Eiffel Tower weigh?
3. How many people visit the Eiffel Tower each year?

A: pg. 213

Eiffel Tower

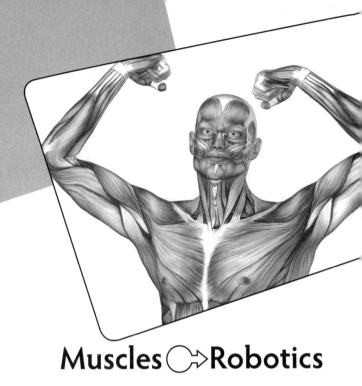

Muscles ⟳ Robotics

The mechanical arms of current robotic devices tend to be stiff and awkward to control. In space, mechanical arms tend to push objects away rather than retrieve them. Our own arms are far different, providing extreme agility and gracefulness in motion and control. Multiple arm muscles permit controlled rotation and flexing around lubricated joints. On a simple level, details of the human body have been copied in laboratories at the University of Texas in Dallas. A wire is connected between pulleys inside a flexible plastic tube. When sections of the wire are heated or cooled, slight expansion or contraction of the length occurs. This loosening or tightening of the wire bends the plastic tube, somewhat similar to our arm muscles.

Temperature-sensitive materials are already used in a host of products ranging from thermostats to medical stents. However, application to robotic or prosthetic limbs raises technology to a higher level. The

stretchable wire can be called an artificial muscle. There are ongoing experiments with heat-generating chemical reactions that can control the wire temperature and its motion. One technique is to generate heat in films of platinum nano-particles that coat the wire. Clearly, it is not easy to duplicate the muscles within our bodies.

Reference

Weiss, Peter. 2006. Pumping alloy. *Science News* 170(1):8–9.

Questions for further study

1. How many muscles are in the human body?
2. Which metals have the largest thermal expansion?
3. How much weight is an adult capable of lifting?

A: pg. 213

Robotics

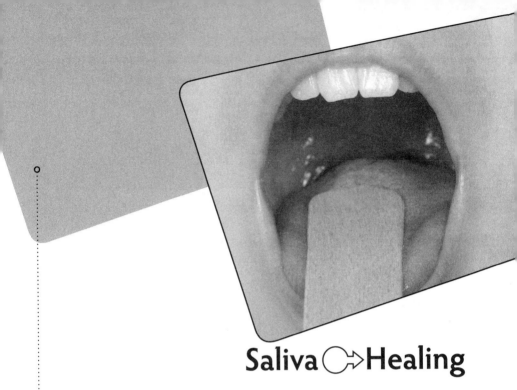

Saliva ↻→Healing

T his example of a practical design in nature may lack appeal, but it is vitally important to our health. A common phrase is to "lick one's own wounds." This saying expresses the effort to care for one's own needs and generally to look after oneself. There is a more practical application, however. Dutch researchers studied the chemical compounds in human saliva. They found an abundance of simple proteins called *histatins* that fight infection. Also present were compounds that cause epithelium skin cells to close over a wound. Most of us have experienced the benefits of these saliva components. Small cuts inside the mouth tend to heal more quickly than external injuries, and mouth healing leaves little scarring.

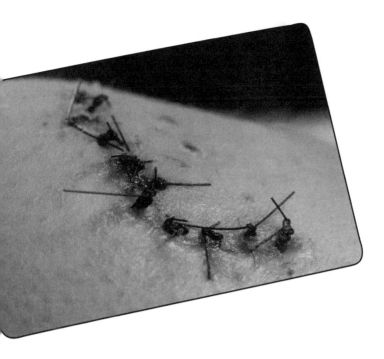

The study of histatins is changing approaches to clinical medicine. It was previously assumed that mouth saliva contained complex compounds called growth factors that aided healing. However, growth factors are not found in significant amounts. A major benefit is that histatins are easier to manufacture and purify than growth factors. Thus, histatins are now emphasized in medical treatment.

Reference

Wright, Karen. 2008. Why wound-licking works. *Discover* January, p. 54.

Questions for further study

1. Would not saliva infect an open wound?
2. Why do animals sometimes lick their wounds?
3. How was the healing property of histatin verified?

A: pg. 214

Healing

Skin ⟳ Self-repairing Plastic

Our skin has the amazing ability to heal itself from abrasions. Suppose the skin's protective outer layer, called the *epidermis*, is injured. When this occurs, the inner dermal layer rushes nutrients to the cut to begin the healing process. Initially, a blood clot forms to cover injured cells, stem the loss of blood, and prevent infection. The repair and regeneration of skin cells begins quickly. The details of the healing process are not well understood, but they are essential to our health.

New materials are under development that mimic the healing ability of skin. Some aircraft components and medical implants are made of strong plastic or polymer. It is possible to prepare this material so that it self-repairs if a crack appears at a later time. A catalyst material is embedded in the outer coating of the polymer surface. The inner region contains a network of small channels filled with a liquid chemical resin.

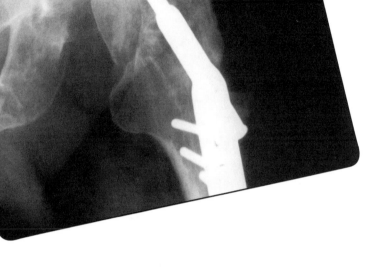

When a crack or fracture reaches the underlying channels, the liquid resin oozes outward and meets the catalyst, and hardening takes place. As a result, the crack is filled and the fracture "heals." This process of mending cracks in plastic can be repeated multiple times, until the internal liquid agent is consumed. Fortunately, our skin repair system is good for a lifetime.

Internet search words:
  self-repairing plastic,
  skin

Questions for further study

1. How many square feet of skin does an adult have?
2. What is plastic surgery?
3. Is artificial skin available in medicine?

A: pg. 214

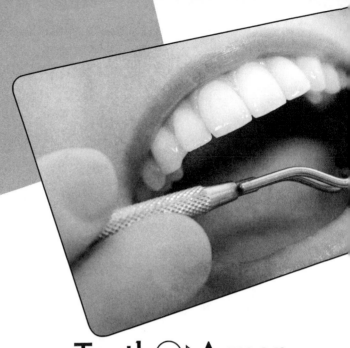

Tooth ⟳→ Armor
Enamel   Coating

Tooth enamel is the strongest material in our bodies. The ceramic-like enamel covers our teeth and is about as thick as a dime. A pliable dentin material beneath the enamel layer acts as a shock absorber to prevent cracking of the tooth. Dentists attempt to mimic the tooth enamel when they apply fillings, crowns, and implants. However, most replacements eventually crack, dislodge, or wear down. Dental researchers are getting closer to making a true duplicate to enamel. Success may

Armor Coating

lead to exact replacements for diseased or damaged teeth. The artificially manufactured enamel will have applications far beyond tooth protection. Many product surfaces would be improved by a coating of true enamel armor.

Reference

Goho, Alexandra. 2005. Something to chew on. *Science News* 167(20):312–313, 316.

Questions for further study

1. What is the composition of tooth enamel?
2. Did George Washington have wooden false teeth?
3. What material is used in making modern artificial teeth?

A: pg. 214

Armor Coating

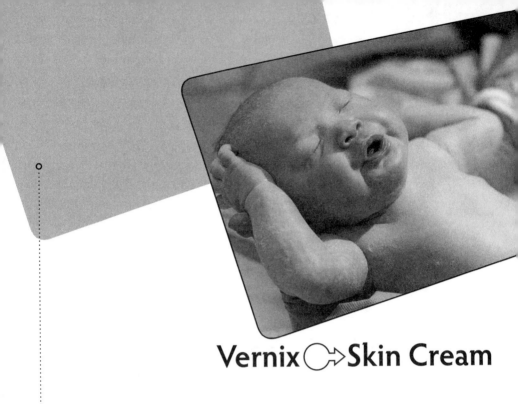

Vernix ⟳ Skin Cream

The skin of an infant before birth is typically covered with a layer of paste-like material called *vernix caseosa*. It is usually reduced to a thin layer by the time birth occurs. This coating helps a newborn in several ways. Vernix provides a barrier to infection, both before and during childbirth. It may also prevent heat loss as the infant enters the outside world. Vernix is 80 percent water, which moisturizes the newly exposed skin. It also contains a high level of vitamin E, which helps the skin deal with possible stress from chemicals and ultraviolet light. Other

roles of vernix are not yet known but may include aid to skin growth. Researchers are attempting to duplicate the chemical composition of vernix. A synthetic form would be useful in treating skin problems for people of all ages. For good reason, vernix is called nature's perfect skin cream.

Reference

Westphal, Sylvia. 2004. The best skin cream you ever wore. *New Scientist* 181(2430):40–41.

Questions for further study

1. What is the origin of the name vernix caseosa?
2. Do animals also produce vernix?
3. What is ultraviolet light?

A: **pg. 215**

7

VEGETATION

The reader may notice that this is a longer chapter of the book. Botany, the study of vegetation, has a rich history of exploration and research. The plants and trees selected here for design insights range from the Venus flytrap to pine trees. The world of vegetation presents us with countless ideas for construction, medicine, and even spacecraft design.

And the earth brought forth grass, and herb yielding seed after his kind, and the tree yielding fruit, whose seed was in itself, after his kind: and God saw that it was good.

— Genesis 1:12

Beech ⟳ Space
Leaf   Antenna

The beech tree begins each growing season with new leaves compactly folded within small buds. When the bud opens, each new leaf unfolds like an accordion. The intricate folding of the beech leaf can be illustrated with a sheet of paper as an exercise in origami, the art of paper folding. Alternate folding patterns occur for leaves from other trees such as hornbeam, and also the wings of beetles and butterflies.

The folding examples from nature have a great range of applications. Consider highway maps that many of us have struggled to refold after use. The maps are easy to open but more difficult to refold correctly. A folding design somewhat similar to the beech leaf shows promise for simplified map use. A single pull extends the map to full size, and a downward push collapses the map once again. Far beyond the challenge of map folding, leaf patterns from nature also find application in space.

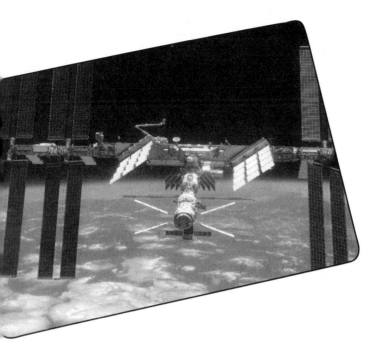

Flat solar panels for satellites require as much surface area as possible to collect solar energy, yet they must be folded compactly for the initial rocket journey into space. Solar panel packaging similar to tree leaf buds was successfully deployed on the Japanese Space Flyer Unit in 1995. Unique new building structures are also modeled after nature's folding patterns.

Reference

Forbes, Peter. 2005. *Gecko's Foot.* W.W. Norton & Company, New York, pp. 181–183.

Questions for further study

1. What is the origin of the word *origami*?
2. How does a butterfly unfold its wings?
3. Is there an example building structure based on origami?

A: pg. 215

Space Antenna

Fully 60 percent of our modern drugs originally come from plants or plant derivatives. As an example, the U.S. National Cancer Institute has identified more than 3,000 plants from which anti-cancer drugs can

Chemicals ⟲⇢Medicine

be made. The medical use of many specific plants has been learned from native societies such as American Indians and Australian aborigines. This valuable field of cultural study is called *ethnobotany*. Such groups, living close to nature, experimented with plants over the centuries to determine which ones are medically effective. Much plant potential still remains undiscovered. In fact, less than one percent of tropical forest plants have been examined for their chemical compounds. Nevertheless, the medical dividends from plant research have been impressive. Table 1 (see pages 148–149) lists some representative plants and their medical treatments.

Some wild plants have been harvested to the point of near extinction. Also, there has been abuse of some other plant products such as the hallucinogenic varieties. There are about 150 known varieties of these mind-altering plants. With care, they are useful in medicine for the relief of pain. For recreational use, however, they impair normal life and are dangerous. Although botanical drugs are a very valuable aid for health, some extravagant claims are not supported by data. It is our responsibility to evaluate natural remedies, sometimes called alternative medicine.

Many chemicals derived from animals likewise have medical

applications, and several examples are included in table 2. Also, many animals have the ability to self-medicate using plant or mineral materials. When they become ill, the creatures know which kind of material to consume for relief. For example, when an African chimpanzee becomes ill in the wild, it may chew on the inner pith of a plant called bitter leaf. Several medicinal chemical compounds are present in this pith. The study of this amazing self-treatment activity is called *zoopharmacognosy*.

Internet search words:
ethnobotany,
medicines, plants

Questions for further study

1. How was aspirin discovered?
2. What actually is aspirin?
3. Does Scripture comment on plant use for healing?

A: **pg. 216**

Medicine

Table 1. A list of representative plants and medical applications. In several cases the derived drugs are shown in parentheses.

Plant	Medical application
Acacia tree	Inflammation (avicins)
Aloe vera	Burn treatment
Araucaria araucana (monkey puzzle)	Dietary supplement
Baobab tree bark	Back pain
Calabar bean (West Africa)	Glaucoma
Camptotheca acuminata (Chinese happy tree)	Tumor treatment (camptothecin)
Catnip	Insect repellant (nepetalactone)
Celery seed	Gout, arthritis
Chinochona bark	Malaria (quinine)
Dandelion	Digestion
Ephedra plant	Respiratory disease (ephedrine)
Foxglove	Heart disease
Fungus (Norway)	Immune system (cyclosporine)
Garlic	Infection (allicin)
Ginger	Antiviral compounds
Goldenseal herb	Antibiotic
Harpagophytum procumbens (devil's claw)	Diabetes, arthritis
Hawthorn shrub	Heart
Jateorhiza palmata vine	Lessens pain
May apple	Cancer (pseudophylum peltatum)
Milk thistle	Liver
Mint oil	Arthritis, nerve damage
Pacific yew	Cancer (taxol)
Peppermint	Stomach
Poppy seed	Insomnia
Prunus africana evergreen	Benign prostatic hyperplasia
Qing Hao herb	Malaria

Plant	Medical application
Rosy periwinkle	Lymphocytic leukemia (vincristine)
Sankerfoot plant (India)	Hypertension (reserpine)
Saw palmetto palmv	Prostate
St. John's wort	Various
Sweet wormwood	Malaria (artemisinin)
Wassabi vegetable	Anti-infection (isothiocyanates)
Water hyssop	Mental illness
White pond lily roots	Cervicitis
Willow tree bark	Aspirin (salicylic acid)
Yams	Steroids (diosgenin hormone)

**Table 2. A list of animals and medical connections.
The drugs in parentheses have been derived from the animal.**

Animal	Medical application
African clawed frog	Produces antibiotics that reduce foot ulcers in diabetics
Bee stings	Induces steroids that reduce arthritis pain
Bioluminescent creatures	Skin disease (coelenterozine)
Cone snail	Neurological disorders (conotoxin peptides)
Coral	Anti-inflammatory (pseudopterosen E)
Giant Israeli scorpion	Brain cancer (chlorotoxins)
Horseshoe crab	Their blood helps detect bacteria
Manuka honey	Staph infections
Mice, rabbits, etc.	Test animals for new drugs
Pigs	Heart valves can be used in humans
Pit-viper	Microbe-fighting myotoxins
Poison-dart frog	Pain killer (epibatidine)
Saw-scaled viper	Anticoagulant (aggrastat)
Southern copperhead	Tumor treatment (contortrostatin protein)
Spider venom	Active compounds used in molecular research
Sponge	Immunosuppressive (discodermalide)
Thailand cobra	Nerve disorders (immunokine)

T his story involves Swiss engineer Georges de Mestral (1907–1990). One day in 1948, de Mestral was hiking in the Jura Mountains with his hunting dog. Following the hike, he noticed cockleburs caught in

Cocklebur ↪ Velcro

his wool hunting pants and also in the dog's fur coat. Cockleburs are round seedpods with a prickly surface that readily attach to clothing or animal fur. Similar examples include burdock and mountain thistle. Close inspection with a magnifier reveals hundreds of tiny grasping hooks on the cockleburs. Based on this finding, de Mestral invented the *velcro* fastener, which usually is made of tiny flexible nylon hooks and loops. The idea was patented in 1951 in Switzerland. The name *velcro* comes from the French words for velvet (*velours*) and hook (*crochet*). It is a corporate name and also a trademark, Velcro®.

The Velcro fastener is used on clothing and shoes to replace zippers, buttons, and snaps. Velcro is also useful today for medical bandages, wall hangers, and countless other fastening applications. Research is underway to produce micrometer-size electromechanical velcro surfaces called "mems." They offer an adhesion ability that can be instantly turned on or off as needed. The "simple" cocklebur has led to a most useful product, Velcro, which is perhaps the best-known example of biomimicry.

Feathers likewise display nature's velcro. Attached to the central hollow spine of a feather are lightweight filaments that overlap and

Cocklebur

150 ↪ Velcro

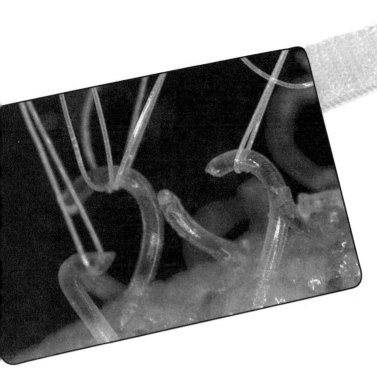

interlock with tiny hooks. The filaments can be pulled apart with an audible snap, and then reattached by stroking the feather, somewhat like a zipper. When birds preen themselves, they reseal their feathers and also waterproof them with a thin coating of oil. The interlocking filaments provide a smooth wing surface for flight, and they also trap air for insulation. Velcro is known as one of the great inventions of the last century. However, it has existed since creation in cockleburs and feathers.

Internet search words: cocklebur, de Mestral, Velcro

Questions for further study

1. What are some novel uses of Velcro?
2. How strong is Velcro?
3. Where is Velcro Valley?

A: pg. 216

Velcro

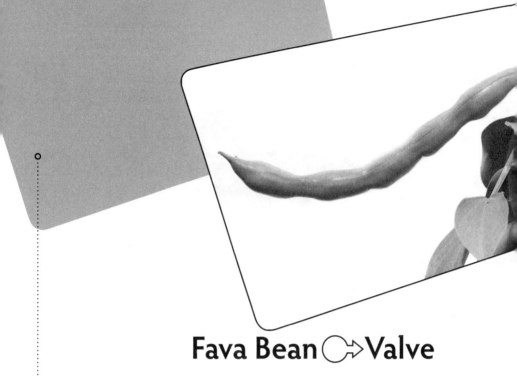

Fava Bean ◯→ Valve

F ava beans are also called broad or horse beans. They were a common food source in early Europe. The bean's internal mechanisms, however, are distinctly modern. Within the capillary tubes of the bean plants are very unusual protein complexes called *forisomes*. These needle-shaped structures are situated in the stem's "sieve tubes" where they control the flow of fluids by either contracting or expanding. Their flexibility appears to depend on the concentration of calcium atoms in the fluid. The result is that internal fluid flow is quickly turned on or off as required by the plant.

Technical applications include fluid control valves in micro devices. The forisomes are a thousand times larger than nanoscale, and thus they fit the important realm of microtechnology. Further uses include micro forceps for grasping objects and pistons for movement in micro motors.

Reference

Ramsayer, Kate. 2003. Dream machines from beans. *Science News*, 164(12):180.

Questions for further study

1. What is the origin of the name *fava*?
2. What are fava beans used for today?
3. What is an unusual use for fava beans?

A: pg. 217

Valve

Fescue ⟳→Herbicide Grass

Homeowners and farmers wage a constant battle against unwanted weeds. Today there is heavy reliance on commercial herbicides, or weed killers. However, nature offers a better solution. Many plants generate chemical defenses against competing plants in a behavior called *allelopathy*. One example of natural weed control is provided by a variety of fescue grass called Intrigue. This grass has good color, grows slowly, resists disease, and inhibits common lawn weeds. The roots of Intrigue produce an amino acid that is toxic to nearby weeds.

Genetic research is extending the allelopathic ability of grass to grain crops, including rice, wheat, corn, and barley. Natural sorghum already

gives off a chemical that discourages competing plants. In addition, some plants manufacture natural pesticides to combat insects. In general, plants tend to produce these defenses only after receiving chemical signals from invasive weeds or insects. Nature has much to teach us about environmentally friendly pest control.

Reference

Raloff, Janet. 2007. Herbal herbicides. *Science News* 171(11):167–169.

Questions for further study

1. What is the most-used herbicide worldwide?
2. What are some other examples of allelopathy?
3. What are some varieties of fescue grass?

A: **pg. 217**

Lotus ⟳→ Surface
Flower　　 Cleaner

T he name *lotus flower* refers to a large variety of plants that grow in water. One distinction is the ability of their large leaves to quickly shed water and dust. Even in muddy water the lotus plant remains completely clean. German scientists discovered the reason for this "lotus effect" in the 1990s. Lotus leaves are covered with countless tiny bumps 0.005–0.01 millimeters high, and also a waxy film. Water droplets, because of their surface tension or stickiness, touch the leaf surface only at the high points. Because of this limited contact area, the water drops quickly roll off the leaf. Along the way, they pick up soil particles like tiny snowballs and pull them off the leaf. As a result, the lotus leaf surface stays clean and relatively dry even during a heavy shower. Surfaces that readily shed water are called *superhydrophobic*. Waxed paper and automobile surfaces are common examples; however, their water-shedding ability is far surpassed by lotus leaves.

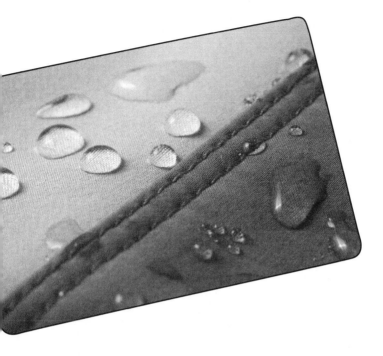

The German chemical company called *BASF* has duplicated the lotus leaf surface with a self-cleaning spray that repels water, dust, and grime. The coating is ideal for rough surfaces such as furniture, clothing, leather shoes, and masonry. The lotus effect has also led to self-cleaning paint that is washed clean with every rainfall. An umbrella that instantly shakes dry is also available, with fabric based on the lotus plant.

Reference

Summers, Adam. April 2006. Secrets of the sacred lotus. *Natural History* 115(3):40–41.

Questions for further study

1. What is the origin of the name *sacred lotus*?
2. What does the seed head of the water lily resemble?
3. What is surface tension?

A: **pg. 217**

Surface Cleaner

Osage ⟳→Barbed
Orange   Wire

Livestock have long been controlled with fences made of stone or wood. However, pioneers on the North American prairies found these materials in short supply. Instead, a thorny bush called Osage Orange, native to East Texas, became popular fence material. The thorn fences worked well, but they were immovable and were a nuisance to grow and maintain.

In 1868, inventor Michael Kelly patented an early form of barbed wire. He wrote, "My invention [gives] to fences of wire a character approximating to that of a thorny hedge." His business was called the

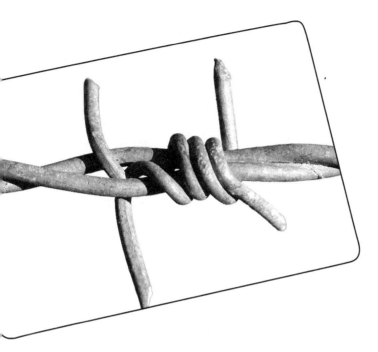

Thorn Wire Hedge Company. Kelly's original barbed wire was modified and improved over the years. Current consumption in the U.S. is more than 100,000 tons of wire each year. This equals a strand of barbed wire well over one million miles long. The Osage Orange thorn bush has led to a major farm industry.

Internet search words:
 barbed wire, Michael
 Kelly, Osage Orange

Questions for further study

1. Does Osage Orange produce edible fruit?
2. Where might one find a "barbed wire museum"?
3. What is razor wire?

A: **pg. 218**

Osage Orange ⬅ 159

Barbed Wire

Pine Cone ↪ Smart Clothes

Smart clothing does not refer to the latest colors and styles available in stores. Instead, they are high-tech clothes that automatically adjust to changing weather conditions. When body temperature increases, tiny openings appear in the fabric to increase air circulation. As temperatures fall, the openings reseal.

The inspiration for smart clothes comes from pine cones. After pine cones drop to the ground, they dry out, open up, and release their internal seeds. The outer scales have two layers of stiff fibers running in different directions. As the fallen cone dries, the outer fiber layer shrinks more than the inner layer, causing the scale to bend outward and expose the inner seed.

The countless tiny flaps that cover smart clothing, function similarly to the pine cone scales. Each flap is only the width of a human hair. The fabric making up the flaps expands and contracts with temperature,

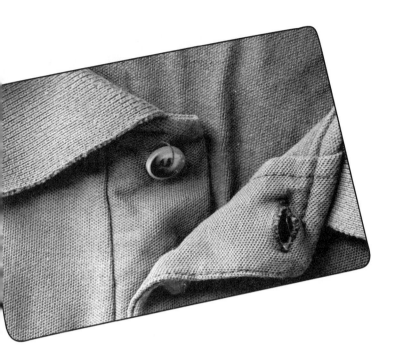

opening and closing as needed. In effect, one wears a host of micro-thermostats. The clothing "opens up" only on the parts of the body where cooling is needed. Smart clothes were first developed in England for military use. Biomimicry expert Julian Vincent is credited with the clothing development. In time, outfits will appear in stores with this built-in air-conditioning ability.

Internet search word:
smart clothes

Questions for further study

1. Are some pine cones opened only by fire?
2. Why do some pine cones open up when warm and dry, and close again when wet and cold?
3. What are some future plans for smart clothes?

A: pg. 218

Rubber ⟳ Automobile
Tree   Tires

On one of Christopher Columbus's several sailing voyages to North America, he noticed natives in Haiti playing games with a flexible ball. Columbus learned that this ball was made by hardening a gummy fluid tapped from certain trees. European visitors soon realized that the material was also useful for making shoes and for rain-proofing fabrics. Around 1770 scientist Joseph Priestley discovered that the tree gum also erased pencil marks, hence its popular name *rubber*. In 1839 Charles Goodyear found that the addition of sulfur greatly improved

the material's strength and durability. This process is called *vulcanization*. Today, a synthetic variety of rubber accounts for 60 percent of the world's production. Plantations also contribute millions of tons of natural rubber each year. Liberia in western Africa and Asia are major producers. From the valuable rubber tree come many thousands of useful products.

Internet search words:
history, rubber

Questions for further study

1. How is tree gum obtained from the rubber tree?
2. Besides pencil erasure, what other great discovery was made by Joseph Priestley?
3. How does sulfur benefit the durability of rubber? A: pg. 219

Automobile Tires

Skunk ⟳ Thermostat Cabbage

Several hundred plant species are able to generate "body heat," somewhat similar to warm-blooded or endothermic animals. Plant examples include the cycad, magnolia, Dutchman's pipe, lotus, water lily, philodendron, and the skunk cabbage. This latter plant has the ability to bloom in early spring inside a snow bank, and may produce a miniature ice cave. Its stalk can reach a temperature that is 63°F or 35°C warmer than the cold surrounding air. In summer, one purpose of the plant's heat generation is the emission of strong odors that attract pollinating insects. The skunk cabbage is well-named. To Native Americans, the skunk cabbage was known as a "famine food." Its roots can be roasted or dried and made into flour.

One potential application of the skunk cabbage is a thermostat for heating and air conditioning. The typical mechanical thermostat, unchanged for decades, uses bimetallic temperature expansion. A metal

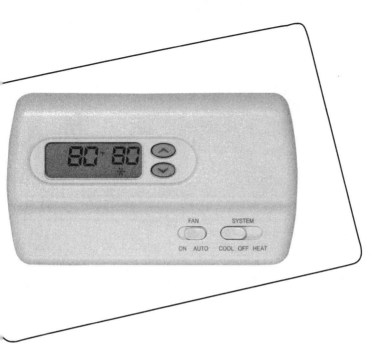

ribbon expands and contracts with temperature, turning the electric circuit on and off. In contrast, the skunk cabbage uses chemical compounds, not yet fully synthesized, to regulate its temperature. This system is sensitive and nearly solid-state in its operation. Japanese scientists have succeeded in regulating an electric heater using a feedback mechanism suggested by the skunk cabbage mechanism. This lowly plant with its questionable fragrance has become a teacher of technology.

Reference

Milius, Susan. 2003. Warm-blooded plants? *New Scientist* 164(24):379–381.

Questions for further study

1. Describe the odor of the skunk cabbage.
2. Besides the skunk cabbage, what other plant is known for its offensive odor?
3. How does a thermostat bimetallic strip function?

A: pg. 219

Thermostat

Spinach ⟳ Solar Cell

Plants collect and use sunlight in the process of photosynthesis. This process converts carbon dioxide and water into new plant material while releasing oxygen. Scientists have tapped into this plant ability in the manufacture of solar cells. First, they harvest light-sensitive proteins from spinach leaves. A layer of this protein material then is deposited on a glass slide with silver electrode wires attached. When light shines on the surface, the spinach proteins generate a weak electrical current.

The protein-based solar cells have several advantages over the more familiar silicon cells. Pure silicon is difficult to produce, expensive, and fragile. In addition, living plants are able to repair or replace degraded proteins. A self-repairing solar cell far outlasts man-made cell varieties. Also, spinach is easily grown and harvested, while the production of silicon solar cells requires high temperatures and an ultra-clean environment. Spinach leaves are helping our efforts to generate free energy provided by the sun.

Spinach

On a more basic level, there are ongoing laboratory efforts to duplicate the photosynthetic process itself. The sciences, called *photochemistry* and *molecular electronics,* seek to understand how plants harvest solar energy. The natural chemical reactions are found to be deeply sophisticated. The goal of this work is the efficient collection and use of light energy.

References

Goho, Alexander. 2004. Protein power. Science News 165(23):355–356.

Gust, Devens and Thomas Moore. 1989. Mimicking photosynthesis. Science 244(4900): 35–41.

Questions for further study

1. What is the chemical equation for photosynthesis?
2. How do silicon solar cells work?
3. What is the potential of solar energy on earth?

A: pg. 220

Solar Cell

T he carnivorous Venus flytrap is well known. Unwary insects are trapped when the plant's cup-shaped leaf snaps shut. The mechanism, called *snap instability*, resembles half of a tennis ball flipping inside out. When open,

Venus ↻ Food Flytrap Packaging

the leaf sides are convex, or rounded. When triggered, the opposing leaf surfaces snap inward, forming a cavity and closing up. There are small hairs or cilia along the edges of the leaves. A slight disturbance of the cilia by an insect triggers the instant collapse of the leaf. The plants are able to distinguish insects from raindrops. Also, the leaf does not close unless two adjacent cilia are moved, or one cilia is touched twice, to prevent false alarms.

Researchers at the University of Massachusetts have fabricated a polymer surface that mimics the flytrap leaf. A flexible, stretched surface is covered with hundreds of small depressions, each about the diameter of a human hair. A similar layer is placed over the first with the bumps upward, making small enclosed pockets. When the top layer is slightly disturbed, its bumps instantly reverse direction, and the layered surface then looks like an egg carton.

The polymer material can be made extremely sensitive to touch, temperature, light, electricity, or chemicals. The possible applications are many. With a clear polymer, the surface resembles an array of lenses.

Venus Flytrap

These lenses can be instantly altered between the convex and concave variety. In food packaging, a change or wrinkle of the polymer surface could result from chemicals indicating food spoilage. In medicine, the enclosed polymer containers could transport drugs in the bloodstream, snapping open when they reach their target area. The Venus flytrap has much to teach us.

Reference

Webb, Sarah. 2007. Snappy transition. *Science News* 172(21):324.

Questions for further study

1. Where do Venus flytrap plants grow?
2. What other plants are carnivorous, or insectivorous, besides the Venus flytrap?
3. Do flytraps ever catch animals larger than insects? A: pg. 220

Food Packaging

Water ⟳ Construction Lily

The first World's Fair took place in London in 1851 and was called the Great Exhibition. A design competition was held for the central display hall. To everyone's surprise, the engineers and architects lost out to a plan prepared by a botanist named Joseph Paxton. It is claimed that his building design was based on the structure of a tropical water lily called *Victoria amazonica*. Paxton had earlier cultivated and studied this lily. The plant displays very large, round floating leaves, some reaching eight feet across with a raised rim. Such lily pads can easily support the weight of a reclining person. Of special interest is the intricate pattern of veins on the underside of the lily leaves. A related pattern of support members was used in Paxton's building, called the Crystal Palace.

The building structure contained more than 200,000 panes of glass, 400 tons total, placed in an iron framework. The glass enclosed both the walls and the roof. The building shell was 108 feet high and covered

18 acres of ground, equivalent to a large modern shopping center.
Critics warned that the completed structure would collapse under its
own weight. Instead, like the Victoria lily, the Crystal Palace was an
outstanding success. It was later disassembled and rebuilt at another
British site, where it stood for 80 years through wind and storm, until it
was destroyed by fire in 1936.

Internet search words:
water lily, Crystal
Palace

Questions for further study

1. What is the origin of the name
   Victoria amazonica?
2. What was inside the Crystal
   Palace?
3. What is the water lily vein
   pattern?

A: pg. 221

Construction

One secret of wheat's success is the ability to plant itself in soil. Wild wheat is dispersed around the world, most notably in Israel. Two fiber strands, called *awns*, extend outward from the seed. As the ripened grain seed falls to

Wild ⟳ Humidity
Wheat Sensor

the ground, the trailing fibers cause the opposite, pointed end to stick into the ground. In dry air, the fibers spread outward and apart. Later, when moistened by dew or rain, the fibers return to their straight, side-by-side position. Over several days, this back-and-forth movement serves to push the seed directly into the soil. Small hair-like barbs along the fibers ensure that the seed can only move downward and not pull loose. When the soil is too hard for penetration, the swimming motion of the fibers slowly pushes the wheat seed along the ground to another location.

The bending motion of the wheat fibers results from unequal expansion and contraction of the sides of the fibers. The result is similar to the action of the metal thermostat found in vehicles and homes. The mechanics of the wheat seed were explored by scientists at Germany's Max Planck Institute of Colloids and Interfaces, in 2007. Scientists suggest that the wheat fiber mechanism may be copied to give motion to micro machines. In addition, a scaled-up version of the wheat mechanism may serve to convert solar energy into energy of motion. This renewable energy supply could then be turned into electricity.

Wild Wheat

Another common wild plant, called *filaree*, displays a self-planting ability in a way distinct from wheat. The seed of this dry-climate plant has a stem with a spiral shape. Moisture changes cause the stem to slowly rotate and drill the seed into the soil, similar to a turning screw. It appears that wild wheat, filaree, and all other plants are programmed for success.

Humidity Sensor

Reference

Elbaum, Rivka, Liron Zaltzman, Ingo Burgert, and Peter Fratzl. 2007. The role of wheat awns in the seed dispersal unit. *Science* 316(5826):884–886.

Questions for further study

1. How do wild and domestic wheat differ?
2. What are some common names for wheat varieties?
3. Where is the filaree plant found?

A: pg. 221

8

NONLIVING OBJECTS

The exploration of useful designs in nature is not limited to the living world. From snowflakes to stars, the intricacies of material objects pay practical dividends to those who look closely. This final chapter goes beyond the popular term biomimicry, the copying of design ideas from living things. One could coin new terms including chemical mimicry (buckyballs, nanoparticles), geo-mimicry (opals), astro-mimicry (pulsars), and hydro-mimicry (water).

> He telleth the number of the stars; he calleth them all by their names
>
> — Psalm 147:4

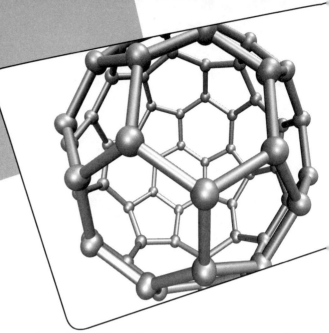

Buckyballs ⤳ Micro Ball Bearings

The graphite and diamond forms of natural carbon have been known from antiquity. In 1985, however, a new variety of carbon was discovered. It has the shape of a sub-microscopic hollow sphere with 60 carbon atoms on its surface. This intricate lattice structure somewhat resembles a soccer ball. These spherical carbon molecules are called fullerenes or "buckyballs." This latter name is given in honor of architect R. Buckminster Fuller (1895–1983), who popularized geodesic domes in his building designs. Since 1990, scientists have been able to make the spherical molecules in large numbers.

Buckyballs are currently used as lubricants, electrical insulators, and surface coatings. A possible future use is ball bearing components for nanotechnology motors and pumps. Buckyball spheres also can carry

molecules placed within them, somewhat like a bird in a cage. Of special interest, medical drugs can be inserted within these spheres. This has the potential for providing timed release of drug dosages.

Another new chemical form of carbon discovered in 1991 is called a *carbon nanotube*. These microscopic cylinders reveal a symmetric arrangement of many carbon atoms on their surfaces. Applications on the microscopic level include containers, conduits, and a template for fabricating other small-scale structures.

Reference

Wu, C. 1996. Buckyballs bounce into Nobel history. *Science News* 150(16):247.

Questions for further study

1. What is the geometric shape on the surface of soccer balls and buckyballs?
2. How does the size of a buckyball compare with the period at the end of a sentence?
3. What is the benefit of an architectural geodesic dome?

A: **pg. 221**

Micro Ball Bearings

Nanoparticles ⟳→Water Purifier

A water treatment plant is a major investment for a city. Impure water passes through a series of filters, solid separators, and chemical treatments before it is released to the environment. These facilities are designed to duplicate the natural purification of water that occurs in streams and rivers.

There remains a desperate need for clean water in many places around the world. It is estimated that water-borne pathogens are responsible for the deaths of at least 15 million children each year. Nano-scale technology offers promising solutions to this problem. Tiny metal particles, the size of the finest dust, can break down poisons without producing harmful by-products. The nanoparticles may be pumped directly into groundwater, spread across the surface of ponds, or applied to oil spills. Iron particles are found to be especially effective as cleaning agents. As the iron rusts or corrodes, water contaminants are converted into harmless compounds. Amazingly, a single handful of the fine dust has a total outside area exceeding nine million square feet. This surface area equals 3,200 tennis

courts. One might say that nanoparticles have much more outside surface than inside space.

Other water treatment studies involve nano-size titanium dioxide (TiO_2) dust grains that break down oil and sewage spills. Also, fiber filters with a coating of aluminum hydroxide (AlOH) are able to trap many bacteria and viruses. These and other inorganic metals may hold the key to safe water on a global scale.

Reference

Lash, Alex. 2003. Got filthy water? *Popular Science* 263(3):46.

Questions for further study

1. Which countries are in special need of clean water?
2. Do some bacteria "eat" oil spills?
3. One nano iron particle, the site of the finest dust, contains how many iron atoms?

A: pg. 222

Water Purifier

O pals are gemstones that are well known for their iridescence. The transparent varieties show brilliant, sparkling colors. The source of these colors is an internal structure of tiny layered silica spheres. Incoming light passes through a number of these rounded layers before reflecting back toward the surface. Particular light colors or wavelengths are selected while others are canceled. The combination of refraction, reflection, and diffraction of light between the layers lead to the colorful flashes from the opal.

Opals ⟳→Photonic Materials

Materials such as opal that selectively reflect distinct colors of light are called *photonic*. Further examples include butterfly wings, shells, and feathers. Photonic materials have stimulated much recent optic research. Colorful, synthetic opals can now be grown in the laboratory. Their use in the manipulation of light is very useful for lasers, fiber optics, holography, waveguides, and lithography. This latter application involves the manufacture of solid-state integrated circuits.

A further emerging field for photonic materials is optical computing. Traditional computers operate with electric current in the form of pulses of moving electrons. Laser light pulses can perform the same function and are ten times faster than electricity with less energy loss due to

heat. Photonic crystals are able to control the direction and behavior of light beams. Also, photonic film may eventually enter food packaging, changing color when the contents begin to spoil. Applied to currency, the film gives protection from counterfeiting attempts.

Reference

Johri, M., Y.A. Ahmed, and T. Bezboruah. 2007. Photonic band gap materials: Technology, applications, and challenges. *Current Science* 92(10):1361–1365.

Questions for further study

1. Where are natural opals found?
2. What is lithography?
3. How can a computer operate using light pulses?

A: pg. 222

A among the great variety of stars in space, pulsars are standouts. They are compact, dense stars with a very rapid spinning motion. Pulsars are also called neutron stars because the entire star is collapsed inward to form a sphere of neutrons. This dense core resembles a giant atomic nucleus 10–20 miles in diameter. The collapse is thought to result from the aging process of certain stars. One spoonful of pulsar "stardust" would easily weigh more than an entire train including all its cars and contents.

Pulsar ⟳ Time Standard

Pulsars display an extremely high rate of rotation, the result of their inward collapse. Similar to a skater who rotates and pulls his or her arms inward, the pulsar spinning motion is concentrated and increased. Astronomer Jocelyn Bell discovered the first pulsar in 1967 while working as a British graduate student. The unusual stars had been predicted to exist some years earlier on the basis of gravity theory. Pulsars are identified by their intense beams of emitted radiation. As a pulsar spins, this radiation flashes outward through space somewhat like a searchlight. The initial pulsar that Jocelyn Bell detected gave off pulses of radio signals. For a short time, it was thought that intelligent signals were possibly being received from outer space, somewhat like Morse code. However, many additional pulsars were eventually found in the heavens, including some that emit visible light flashes similar to a strobe light. Over the years more than 1,000 pulsars have been located in our own Milky Way Galaxy.

Pulsar flash times vary from once every few seconds downward to just milliseconds. This latter figure means that some pulsars spin at an

incredible rate of more than 1,000 times per second. The flash rate for each pulsar is very regular, varying by only one part in ten billion. This means that pulsar stars offer one of the most accurate timekeeping systems available, rivaling atomic clocks. Another related type of dense star called a white dwarf also provides an ultra-stable clock by its regular emission of light pulses. One is reminded of the special purpose of stars as recorded in Genesis 1:14, to serve as time keepers. Over several years, pulsars slightly decrease their flash frequency. This gradual slowdown in motion results from the gradual loss of pulsar energy. The loss does not hinder the usefulness of pulsars as precise time standards.

References

Cowen, Ron. 2003. Stellar speed limit. *Science News* 164(18):125.

Roth, Joshua and Alan MacRobert. 2004. White dwarfs as ultrast-able clocks. *Sky and Telescope* 107(2):17.

Questions for further study

1. What is the typical density of a neutron star?
2. How close to earth is the nearest known pulsar?
3. What would happen if we visited a neutron star?

A: pg. 223

Time Standard

Water ⟳ Impeller Flow

There are many similar spiral shapes found in nature, both large and small. These include galaxies, hurricanes, tornadoes, rising smoke, seashells, flowers, and water whirlpools. This last example is of special interest to a California company called PAX Scientific, founded by Australian naturalist Jay Harmon. Jay noticed that fluids, whether liquid or gas, tended to swirl as they moved about. Water demonstrates this circular pattern as it flows down the drain in a sink.

Detailed studies of this twisting, spiral motion show that it is a very efficient way to transport fluids from place to place. The rotating pattern is different from the usual straight-line flow of water passing through pipes or the airflow through ductwork in homes. PAX engineers have designed impeller blades that intentionally produce whirlpools of moving air or water. The new devices outperform conventional blades in

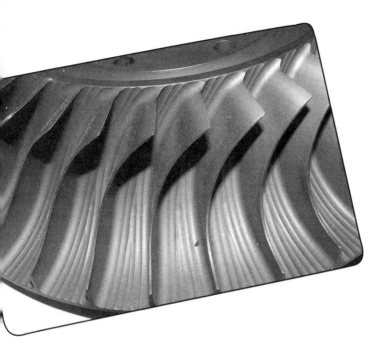

the quantity of fluid moved, and far less energy is used. One of the new, successful blade shapes is modeled on the delicate spiral blossom of the calla lily flower.

The newly designed impeller blades are a key component of fans, pumps, turbines, and propellers. Small, ultra-efficient cooling fans are planned for many devices, ranging from computers to refrigerators.

Reference

Watters, Ethan. 2007. Product design, nature's way. *Business 2.0* 8(5):86–92.

Questions for further study

1. Is there a mathematical formula for the spiral designs found in nature?
2. How does an impeller differ from a propeller?
3. What is the largest spiral observed in nature?

A: pg. 223

Impeller

CONCLUSION

This book has considered many examples of practical design ideas found in nature. They come from both living and nonliving components of the world. Plants, animals, and people — all are included. Engineers and scientists increasingly look to nature for solutions to their complex problems. The space agency NASA studies insects for the design of efficient robotics and spacecraft. Medical research learns from self-healing mechanisms in nature. Answers to environmental questions are found in efficient ecosystems worldwide.

A basic question arises: Why is nature so useful? Some will suggest that evolutionary progress over eons of time has selected optimum design while "weeding out" inferiority. However, the fossil record displays endless examples of exquisite design that are equal or greater than those of the present day. We believe that design in nature reveals the fingerprint of our Creator.

The exploration of practical intelligent design in nature offers a new paradigm for science inquiry. This new direction is the searching out of practical features in nature, and acknowledging their origin by the Creator. The approach is somewhat like unwrapping gifts and using the contents to better mankind. We have indeed been provided with countless practical gifts in nature. Some are obvious, like Velcro. Others require careful unwrapping and study, such as the microscopic world.

It is readily acknowledged that our present world is not perfect. Examples of imperfections do indeed exist in nature. This is a theological issue that began with the Fall of mankind in Genesis 3. Even in a fallen world, however, countless blessings from God's creation surround us. Proverbs 25:2 states, "It is the glory of God to conceal a matter; to search out a matter is the glory of kings." It is indeed an honor and privilege to explore nature and search out the secrets of the Creator. Surely, the great majority of useful designs in creation remain undiscovered. May this book encourage the next generation of explorers.

The website *DiscoveryofDesign.com* catalogues additional design examples. Join our effort and let the authors know of additional useful designs present in the wonderful creation all around us.

ANSWERS TO CHAPTER QUESTIONS

Chapter 1 — Microorganisms

BACTERIA ➔ MICRO-MOTOR

1. What is the precise meaning of the words *micro* and *nano*?

    The prefixes *micro* and *nano* come from Greek roots meaning respectively "small" and "dwarf." Technically, micro stands for one millionth, or 10^{-6}. The, thickness of a sheet of paper is about 4000 micro-inches. A nano is defined as one billionth, or 10^{-9}. In one inch there are one million micro-inches, and one billion nano-inches.

2. How does the speed of an electric fan compare with the 100,000 rpm rate of the molecular motor?

    Typical speed for a household electric fan is 1500 rpm. This is 60 times slower than the speed of the bacteria flagellum.

3. What are the chemical properties of silly string?

    Silly string was first introduced as a child's toy in 1972. A liquid polymer in a pressurized aerosol container quickly turns solid when exposed to air. Polymers are compounds with long chains of chemically bonded molecules. One container can produce silly string hundreds of feet long.

Bacteria ➜ Battery

1. What actually is a battery?

   A battery is a chemical cell useful for storing electrical energy. Energy storage is a challenge for technology, and research continues on batteries with high efficiency and capacity.

2. Why are most energy conversion processes inefficient?

   When fuel is burned in a car to produce motion, much of the resulting heat energy is unused. This heat radiates outward from the engine and water coolant, and also leaves with the hot exhaust gases. The Second Law of Thermodynamics describes such inevitable losses in every energy transfer process.

3. How many electrons pass through a standard 60-watt light bulb in one second?

   A light bulb, whether in a flashlight or reading lamp, has an electric current of about one ampere. This amounts to over six million-trillion electrons (actually 6.25×10^{18}) passing through the bulb filament each second. These electrons move through the filament at a snail's pace, somewhat similar to a large crowd passing through a narrow gate.

Biofilm ➜ Bacteria Control

1. Estimate the number of bacteria on your hands.

   Living bacteria are everywhere in large numbers, and there may well be millions on each of our hands. Most are harmless "resident bacteria," but hand washing is a good practice to prevent the spread of less friendly bacteria. Many bacteria reproduce and multiply in less than a single hour.

2. Where might one find freshwater biofilms?

   A layer of biofilm often coats rocks in streams, making their surfaces slippery. Also, biofilms sometimes cover the surface of stagnant ponds.

3. What are some unusual locations of biofilms?

   Colorful biofilms are found on the surface of hot, acidic pools in Yellowstone National Park, as well as on glaciers. In homes, biofilm colonies may grow in the corners of shower stalls.

Diatom ➜ Nanotechnology

1. Are diatoms plants or animals?

   Diatoms have been variously classified as plants, animals, or something in between. They share biochemical features of both plants and animals, including photosynthesis and mobility.

2. What is the mineral name for glass?

   Glass consists of the chemical compound silicon dioxide, SiO_2. The mineral name is quartz or silica.

3. Diatomaceous earth is a powdered form of diatom fossils. What are some of its uses?

   This common chalk-like material serves as a filter for liquids and an abrasive. Its absorbent property is also useful as a component in kitty litter. In addition, diatomaceous earth serves as a stabilizer in explosives, as a mechanical insecticide, and as a medium for potted plants.

Protein ➜ Solar cells

1. How is electric current measured?

   Current is the flow of electrons through a conductor, measured in amperes. A typical light bulb may use one amp of current. The current produced by a single spinach protein is on the order of a billionth of an ampere, or one nanoamp (10^{-9} amp).

2. How is it possible that wind power, water power, and fossil fuels are all forms of solar energy?

   Solar heating of portions of the earth causes breezes to blow. Wind then results as air moves to even out temperature and pressure differences. The sun also causes water to evaporate from lakes. This water later condenses and flows back downhill, where it may be harnessed by generators for hydroelectricity. Fossil fuels include coal, oil, and natural gas. It is thought that these fuels largely formed from earlier plants that captured sunlight by photosynthesis. The vegetation later was buried and compressed by pressure and heat. Fossil fuels are thus solar energy stored up from the past.

3. Can you name three nonsolar forms of energy?

   Nuclear energy is available from elements inside the earth, and geothermal energy comes from underground magma. Tidal movement, caused by the moon's gravity, is also tapped as an energy source.

Chapter 2 — The Insect World

ANTS ➜ AIRLINES

1. How many legs does an ant have?

    Like all insects, ants have three pairs of legs, totaling six. Myrmecology is the scientific study of ants, a branch of entomology, or insect science. The Greek word for ant is *myrmex*.

2. How does the total number of ants compare with other animals?

    Ants are the most numerous animal worldwide. They comprise 15–25 percent of the total animal biomass or weight. The combined weight of all living ants far outweighs earth's total human population.

3. What did King Solomon say about ants?

    Proverbs 6:6–8 describes the diligent work of the ant. This insect provides us with an example of ambition and planning for the future.

ASIAN BEETLE ➜ PAPER WHITENER

1. What is currently used to give paper a white color?

    Bleach is added in an early stage of the paper-making process. Later, various white pigments may also be added, including the chemical titanium oxide, TiO_2.

2. What is chitin?

    Chitin (KI-tin) is the common shell-like covering of many insects, and is also found on crustaceans such as lobsters. Chitin is a long-chain polymer with the chemical formula $(C_8H_{13}O_5N)_n$.

3. What could be the purpose of the Cyphochilus beetle's white appearance?

    To avoid predators, the color of this beetle matches its frequent habitat, bright white fungi. How the fungi itself produces this color is yet another marvel that is not well understood. The beetle's reflective surface also provides cooling for its body.

BOMBARDIER BEETLE ➜ GAS TURBINE ENGINE

1. Hydrogen peroxide is found in many medicine cabinets. What is its use?

    Hydrogen peroxide, H_2O_2, is useful for cleaning and sterilizing wounds. This chemical is also a disinfectant and bleaching agent.

2. Where are bombardier beetles found?

Bombardiers and similar beetles are widespread across North America, especially in the Southeast states. The beetles inhabit woodlands and often live beneath leaves or stones.

3. How would an evolutionist attempt to explain the origin of the bombardier beetle?

"Just so" stories are suggested whereby the beetle's defense mechanism developed slowly over many generations. In truth, the beetle is an example of irreducible complexity in nature.

BUTTERFLY ➜ COSMETICS

1. How many different species of butterflies have been catalogued?

Estimates total about 20,000 known species of butterfly. In many places, one hundred or more butterfly species are common, a highly unusual variety to coexist in one location.

2. How long does a butterfly live?

Adult butterflies typically live for about one month.

3. What causes the different colors of light?

The colors of light result from their different wavelengths. Wavelengths for visible light are very small, ranging between 0.7 and 0.4 microns for the colors red and blue. These wavelengths are about 0.0001 of an inch long, 100 times less than the thickness of a sheet of paper.

DRAGONFLY ➜ SURVEILLANCE

1. What becomes of dragonflies in wintertime?

Dragonflies may live one to five years, most of it in the larvae stage. The larvae spend winters hibernating beneath the surface of lakes and ponds. The adult dragonfly lifetime is just a few months long and does not carry over a cold winter.

2. Have fossil dragonflies been found?

Dragonfly fossils are indeed found in many locations. The impressions in rock are close copies of living dragonflies, except that some fossils are ten times larger. The impressive fossil wingspans reach 30 inches (76 cm). Other than a reduction in size, dragonflies have not changed noticeably over time.

3. What do dragonflies eat?

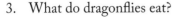

Dragonflies eat other insects, including flies, bees, and midges. They are sometimes called beneficial "mosquito hawks" because they help keep these pests in check.

Firefly ➜ Light Stick

1. What is the lifespan of the firefly?

   An adult firefly lives only about one week.

2. What is the origin of the chemical names *luciferin* and *luciferase*?

   Both words come from Latin, "to bear light." The name Lucifer appears in Isaiah 14:12 as an indirect reference to Satan, who misrepresents himself as an angel of light according to 2 Corinthians 11:14.

3. Why does a group of fireflies sometimes blink off and on simultaneously?

   Suggestions for the flash patterns include communication about food sources, and also social interaction including courtship. The male usually flies above the ground and the female flashes a response from below, showing her location.

Fly ➜ Hearing Aid

1. How may a cricket be used as a thermometer?

   The chirping rate of a cricket increases with temperature. Count the number of chirps in 15 seconds, then add the number 39 to find the Fahrenheit temperature. If 30 chirps are heard in 15 seconds (two per second), then the temperature is about 69°F (20.5°C).

2. How does hearing sensitivity compare with our other senses?

   Of our five senses — hearing, sight, smell, taste, and touch – hearing may be the most sensitive. A loudness level very close to zero decibels is noticeable. At this threshold of hearing, our eardrum moves only about the distance of a single atom. In addition, our sense of hearing is very rapid in its response. One can determine the direction of sound by which ear hears it first, a time difference of about 0.0004 seconds.

3. Does a fly have eardrums?

   Yes, and unlike people, the eardrums of a fly are connected. A sound disturbance that impacts one eardrum quickly moves to the other, and from this the fly's brain is able to determine direction to the sound.

Honeybee ➜ Surveillance

1. What image does an insect see with its compound eyes?

    Each micro lens in a bee's eye has a narrow field of vision, somewhat like a flashlight beam. The partial images from the thousands of lenses blend together in an overlapping picture mosaic. Detail may be lacking, but any slight movement is readily detected. The color vision of bees lacks sensitivity to the color red, but they see more violet and ultraviolet light than the human eye.

2. How small is one of the multiple lenses of the honeybee eye?

    A honey bee has about 7,000 separate lenses in an eye less than one millimeter across. This means that a single lens is about a micron in size, less than the thickness of a sheet of paper.

3. What is ultraviolet light?

    Ultraviolet light, or UV, has a wavelength less than that of visible light. It is also called black light and is more energetic than the familiar visible colors. Ultraviolet light can be used to sterilize medical instruments.

Insect Hearing ➜ Atomic Force Microscope

1. Who invented the Atomic Force Microscope?

    The AFM was developed in 1986 by scientists Calvin Quate and Christoph Gerber. Both men have won many awards for microscope advancements.

2. How small an object can an AFM clearly "see"?

    The AFM "feels"'surfaces rather than sees them. It resolves objects down to one angstrom in size, or 10^{-10} meter. This is the size of an individual atom, and the AMF shows the arrangement of individual atoms in crystal structures such as salt, sodium chloride.

3. List some of the different types of microscopes.

    There are several varieties of microscopes designed for various purposes. The earliest is the optical microscope. It uses visible light and resolves objects as small as 0.00001 inch, or 0.4 microns. Electron microscopes use a beam of electrons to produce three-dimensional images of objects as small as a few angstroms, slightly larger than an atom. Other instruments include the stereo microscope, atomic mass microscope, and the scanning tunneling microscope.

INSECTS → ROBOTICS

1. What is the origin of the word *robot*?

   This word was coined by the Czech writer Karel Čapek in his play *Rossum's Universal Robots* (1923). The root refers to servitude or forced labor.

2. Can one obtain an advanced degree in robotics?

   Many schools offer robotics training, ranging from a two-year technical degree, on through the doctoral level. Specialties include robotic design, assembly, programming, and management. There are opportunities in robotic manufacturing, education, and product development.

3. How many legs do centipedes have?

   The name is French for "one hundred feet"; however, centipedes are not limited to this number. Instead, two legs appear on each body segment behind the head. The number of segments varies between 15 and 177, and is always an odd number. These numbers result in total leg numbers ranging from 28 to 352.

NAMIB BEETLE → WATER COLLECTOR

1. What causes fog or dew to appear in the morning?

   All air holds moisture, even in a desert climate. When the temperature drops below the dew point, often in early morning, the air can no longer hold the moisture in suspension. Dew then precipitates from the air, forming droplets on cool surfaces.

2. What defines a desert climate?

   The standard definition is a region with less than ten inches (25.4 cm) of precipitation annually. Harsh desert regions in the country of Namibia receive ten times less than this, just one inch of rain per year on average. Deserts are not always warm; frigid Antarctica is the world's largest desert because of its sparse snowfall.

3. How do Namib beetles communicate?

   They tap the sand with their tail ends to attract mates. This drumming can be heard by a nearby person, and human tapping can also attract beetles.

Spider Silk → Fiber Optics

1. What was an early use of spider silk by astronomers?

   Many telescopes use a small, low-magnification finder scope to locate distant objects. Typically there are cross hairs in the view of the finder that center on the desired object. In earlier years, these cross hairs were made from spider silk that was attached to a glass lens surface. Spider silk was the thinnest permanent material readily available for this purpose. Telescopes today often use laser light or computer tracking for sighting objects.

2. What is a nanotube?

   A nanotube is a wire-like structure that is usually made of a single layer of carbon atoms. The strands are 50,000 times thinner than the width of a human hair. They can extend several millimeters in length and have many uses in technology.

3. How can it be said that spider silk is stronger than steel?

   Suppose a steel wire is stretched to the same diameter as spider silk. The steel strand then is several times thinner than steel wool fiber. In this extreme limit of thinness, spider silk will support five times the weight of the equivalent steel fiber.

Termite Mound → Ventilation

1. How many termites may live in a single African mound?

   Population studies of the termite mounds show as many as 15,000 adult occupants.

2. Describe how the termite mound heating and cooling vents operate.

   The termites constantly dig new tunnels and vents while plugging others with soil. In outlying areas of Africa, this grainy, reworked mound material is sometimes used for hard-packed floors.

3. What is the average temperature of Zimbabwe's capital city, Harare?

   Average annual temperature varies between 70°F (21°C) in summer and 57°F (14°C) in winter. The city of Harare is located 9° north of the equator.

Timber Beetle Larva → Chainsaw

1. How were big trees harvested before chainsaws were invented?

   Along with the axe, crosscut saws date back to Roman times. These saws cut horizontally through the tree trunk, across the grain. Such saws improved over time with new metal alloys and tooth design. They are still much used around the world, often with a woodsman on each end of the saw.

2. Are timber beetles a major pest in forestry?

There are a large number of beetles, borers, and worms that damage wood. The timber beetle tends to prefer diseased trees, and is also attracted to cut logs. The best defense is healthy trees that repel and withstand invasive beetles.

3. What is the largest commercial chainsaw manufactured today?

Typical chainsaws for home use have a bar 14–20 inches long. Commercial forestry saws have bars 36 inches or longer. All chainsaws are dangerous to operate; the larger saws are powerful and heavy.

WASP ➔ PAPER

1. What is the shape of the paper wasp nest?

The water-resistant nest consists of a gray or brown papery material. There are open hexagonal tubes for rearing the young, usually pointing downward. Small nests may be constructed in the eaves or corners of houses.

2. What is lignin?

This complex chemical compound comprises about one-third of dry wood. Lignin is a major part of the cell walls of plants. The word comes from the Latin for wood.

3. Is some paper still made from linen cloth?

Fine linen paper is made of cotton or linen fabric, beaten into fibers. Certain paper currency is also cloth-based.

WATER STRIDER ➔ WATER REPELLANT

1. How do water striders move from one pond or puddle to another?

Adult striders have wings that permit flight to nearby ponds.

2. How is it that a water strider, or even a needle, can float on water?

Surface tension describes the "stickiness" of liquids. This is the tendency of molecules to adhere together and resist penetration. Water has a very high surface tension, allowing light objects to float. Liquid mercury also has a high surface tension.

3. How do water striders move so rapidly on water?

The striders move by "digging" their feet into the water surface and generating vortices or tiny swirls. Pushing against the resulting "mini wall" of water, they recoil forward rapidly with speeds measured at five feet per second (1.5 meter/sec).

Chapter 3 — Flight

Bats ➜ Sonar Systems

1. What actually is an ultrasound wave, as produced by bats?

Sound is defined as a vibration, whether produced by our vocal cords, violin strings, or thunder. In air, sound is a slight, rapidly changing pressure that is transmitted between the closely packed air molecules. The speed of sound in air depends on temperature and averages about 1,100 feet per second, or 750 miles per hour (335 m/sec).

2. What is the frequency of a dog whistle?

Also called a silent whistle, this device produces a sound frequency between 16,000 and 22,000 hertz. This frequency is at the top or somewhat beyond our human hearing range. Dogs and cats have small ear drums that can vibrate very rapidly and detect such sounds.

3. What is a flying fox?

This name is given to large bats with wingspans reaching 5 feet, or 1.5 meters. They live in tropical areas and are also called fruit bats because of their diet. Some flying fox species use echolocation, while others have large eyes and excellent night vision.

Bird Flight ➜ Aircraft

1. What are the fastest speeds for animals in the air and under water?

Swifts and peregrine falcons are the fastest known creatures with speeds approaching 200 miles per hour. Homing pigeons are also record-setters with speeds variously reported at 100–200 mph. Even the tiny hummingbird achieves an impressive 60 mph. On land, the cheetah manages short bursts of 70 mph. Pronghorn sheep and jackrabbits have speeds slightly less. In contrast, the fastest human runners reach 20–25 mph. Under water, the speed of sailfish has been measured at 70 mph, tuna at 48 mph, and dolphins at 30 mph. Olympic swimmers can reach nearly 5 mph for short stretches. The maximum speed of military submarines is classified, but may exceed 40 mph (35 knots).

2. Give a numerical example for the bird flight formula.

The Eurasian Kestrel is a variety of falcon. The flapping of its wing is measured at 5.61 strokes/second, with an amplitude or maximum wing displacement of 0.339 meters. It's typical speed is 18 mph (8.1 meters/second). The formula gives (5.61 strokes/sec) x (0.339 meters)/(8.1 meters/sec) = 0.235 for the Strouhal number. This low number implies that the Kestral flies with extreme energy efficiency.

3. Do flying fish have wings?

No, instead they have large pectoral fins that enable them to take short gliding flights. In escaping predators, they can glide 164 feet (50 m) or more through the air.

FLYING REPTILE ➜ DELTA WING AIRCRAFT

1. What is the meaning of the name *Sharovipteryx mirabilis* given to the fossil flying reptile?

The first word means "Sharov's wing," named for researcher A. G. Sharov, who reported the Russian fossil discovery in 1971. The second word is Latin for "wonderful," or "miracle," well-chosen for this expert flyer.

2. What is a wind tunnel?

A wind tunnel is a laboratory chamber used to study the motion of animals or vehicles through air. With the object held firmly in place, air is rapidly released through the chamber at high speed. A visible vapor such as smoke can be added to the air flow and photographed as it moves around the object. This vapor indicates streamlined motion of the object in air, or turbulent resistance to motion.

3. What year did the Wright brothers succeed with their flight?

Orville and Wilbur Wright succeeded with the first controlled, powered flight on December 17, 1903. The initial flight lasted 12 seconds and covered a distance of 120 feet. This event took place at Kitty Hawk, North Carolina.

KINGFISHER ➜ BULLET TRAIN

1. Where do kingfishers live?

Kingfishers live worldwide in woodlands and wetlands. About 90 species have been identified with great variety in size and color.

2. How do kingfishers see while under water?

The kingfisher eye has an egg-shaped lens that flexes to focus in both air and water environments. Our eye lens has a similar, more limited, ability to change its focus, called accommodation.

3. What is a sonic boom caused by the bullet trains?

When compressed air rapidly expands, a pressure wave moves outward at the speed of sound. Air molecules transmit the disturbance somewhat like a row of railroad cars bumping into each other. When the disturbance reaches our ear, the eardrum vibrates and we hear the sound. Thunder creates a similar result because air expands rapidly when heated by lightning.

Owl Wing → Noise Reduction

1. What is the largest owl?

Two large examples are the Eurasian Eagle Owl and the Great Gray Owl. Their length from head to tail reaches 28–33 inches. In contrast, two of the smallest owls are the Least Pygmy and Elf Owls, each just 3–4 inches long.

2. How many species of owl have been catalogued?

About 200 species of owls are known. An additional 40 extinct species have been found as fossil remains.

3. How is the loudness of sound measured?

The decibel (dB) measures the loudness of sound, named for Alexander Graham Bell. The logarithmic nature of decibels allows a great range of sounds to be expressed in convenient numbers. Typical loudness values include a quiet room (20 dB), ordinary conversation (60 dB), a truck ten feet away (90 dB), and the threshold of pain (120 dB).

Swift → Aircraft Wings

1. How does evolution explain the origin of flight?

There are at least three leading ideas on animal flight. Each assumes the development of feathers from reptilian scales, and wings arising from forearms. The curosial and aboreal models were proposed a century ago. The former suggests that short jumps became longer and longer over time, until the animal left the ground completely. The aboreal model suggests that creatures glided down from trees, eventually conquering the air. A third model, called pouncing proavis, is more recent. The suggestion is that predators pounced on prey from tree branches. In time the pounce became a swoop, and finally turned into controlled flight. In contrast to such models, the idea of flight created by God from the beginning of creation is the credible and refreshing preference.

2. Since swifts do not perch on tree branches, where do they build their nests?

    Swifts tend to build nests from sticks and mud on vertical surfaces. The locations include caves and chimneys. Swifts also construct nests under the eaves and outdoor decks of homes.

3. Are swifts and swallows the same?

    They may look similar but are quite distinct. Swallows are song birds and tend to fly closer to the ground than swifts.

TOUCAN BEAK ➜ SHOCK ABSORBER

1. In which countries might you find toucans?

    Toucans are distributed across Central and South America. The main country locations include Argentina, Bolivia, Brazil, and Paraguay.

2. Why are toucans so colorful?

    The standard evolutionary reason for the bright and bold male colors found in nature is for advantage in attracting a mate. In the creation view, the colors found everywhere in nature, whether bird feathers or sunsets, show the Creator's glorious artwork.

3. How can the beak of a woodpecker survive hammering?

    A woodpecker can drill into hard wood with 1000 taps per minute. Several safety mechanisms are in place. Its strong bill is separated from the skull with sponge-like cartilage that serves as a shock absorber. A thick skull with spongy bones also cushions the bird's brain. Strong neck muscles keep the head aligned and prevent harmful twisting. The feet have X-shaped toes positioned both forward and backward for firm grasping. Stiff tail feathers provide additional leverage against the tree. There are surely other unknown internal mechanisms that protect the woodpecker. Toucans, woodpeckers, and all other birds are excellent examples of creative design.

Chapter 4 — Underwater Life

BOXFISH ➜ AUTOMOBILE DESIGN

1. How large is the boxfish?

    One variety of boxfish, called the cowfish, reaches a length of 20 inches (50 cm). Most adult boxfish are smaller, around 5 inches (13 cm) long. Small boxfish specimens, just one inch long, may be purchased from pet shops for saltwater aquariums.

2. Where is the boxfish found?

   The boxfish inhabits the Atlantic, Indian, and Pacific Oceans.

3. How many distinct species of boxfish have been catalogued?

   Boxfish are part of a larger order of fish called Tetraodontiformes. This includes 350 known groups such as boxfish, puffers, and ocean sunfish. There are more than a dozen known boxfish species, including the buffalo trunkfish and the long-horned cowfish. Additional species will surely be found.

BRITTLESTAR ➜ MICROLENS

1. Where in the oceans are brittlestars found?

   They live in all the oceans and extend from polar regions to the tropics. Brittlestars thrive in shallow areas, and also are abundant in deeper ocean depths, a mile or more down.

2. What does evolution theory suggest about the origin of vision?

   The great variety of eye structures found today in nature is thought to have developed from a single light-sensitive cell. This original photoreceptor then somehow depressed into a cup or eyeball shape, filled with fluid, and later diverged into the many types of eyes found today. Many scientists prefer the majestic statement of Psalm 94:9, "He that formed the eye, shall he not see?"

3. Besides the lenses of brittlestars, where else might one find the mineral calcite?

   Calcite is one of the most abundant minerals on earth. In addition to the lenses of extinct trilobites and living brittlestars, calcite comprises most seashells. Calcite also precipitates from water to make limestone, and forms large crystals in caves.

CUTTLEFISH ➜ CAMOUFLAGE

1. Name some other animals that change color.

   Along with the well-known chameleon is a host of other color-changing creatures. In water, octopuses and flounders alter their appearance to blend in with surroundings. Frogs, toads, crabs, and prawns also can lighten or darken their skin. The male goldfinch is bright yellow in summer and olive colored in winter. Seasonal color changes also occur for the arctic fox, ermine, and snowshoe hare. It is even suggested by some that dinosaur skin had color-changing features.

2. Are cuttlefish used as seafood?

   Cuttlefish are indeed prized for food, especially in Asian countries. Cuttlefish bones are also softened and eaten. Ongoing genetic research seeks to increase the nutritional value of cuttlefish meat.

3. What is polarized light?

   Light consists of vibrating electric and magnetic waves or fields. In unpolarized light, these moving waves vibrate in all possible directions. When the waves are all aligned in the same direction or plane, the light is said to be polarized. If these waves could be seen within a beam of light, they might all be moving up and down vertically, with no horizontal wave motion.

ELEPHANT NOSE FISH ➔ ELECTRIC SENSOR

1. What is an electric field?

   An electric field is a concept used to describe the invisible force of attraction or repulsion acting at a distance between electric charges. Electric and magnetic fields were first suggested as a visual aid by Michael Faraday (1791–1867).

2. Does an electric eel also generate an electric field?

   Electric eels have several specialized organs that become electrically charged in series, somewhat like the plates of a car battery. The eel's prey may be stunned with 500 volts and one ampere of current. Intense electric fields accompany the eel's electric discharge.

3. What large sea animal has a nose protrusion somewhat similar to the elephant nose fish?

   The narwhale lives in cold Arctic waters and grows to a length of 26 feet (8 m). This creature has a long tusk that extends about half its body length. A spiral groove covers the tusk surface. The function of this tusk is not known, and may be an antenna for echolocation.

FISH MOTION ➔ SHIP PROPULSION

1. How does the speed of Olympic swimmers compare with fish?

   Some fast animal swimmers include the sailfish (68 mph), Mako shark (60 mph), blue fin tuna (43 mph), and dolphins (37 mph). Top human swimmers move through water at about 5 mph.

2. Is it known why whales breach, or leap above the water?

   Humpback whales, especially, leap entirely out of the water, twist, and then land with a loud splash. Suggestions for this activity include communication, defense, skin care, predatory behavior, and simply looking around. Many whale watchers believe that whales breach for the entertainment of spectators.

3. What is the supposed evolutionary origin of whales?

The current view is that certain small land mammals began spending more time in the sea, and their descendents gradually became whales. The mammal's legs changed into flippers, the tail broadened into flukes, and in the buoyant water world the body became enormous. Fossil evidence for these changes is lacking.

LOBSTER EYE ➜ TELESCOPE LENS

1. What is an x-ray?

X-rays are a high-energy form of light. The sun's spectrum of light includes a small component of x-rays. They have a wavelength hundreds of times smaller than visible light and readily penetrate matter.

2. What do x-ray telescopes see?

X-rays are emitted by matter that experiences extreme heat or motion. Thousands of x-ray sources are detected in space. Many of them are at locations where matter is being pulled inward by the gravity of nearby massive stars.

3. Is a crayfish eye similar to that of a lobster?

Yes, freshwater crayfish, also called crawfish or crawdads, have a reflective optics system closely similar to lobsters.

MUSSELS ➜ ADHESIVE

1. Why are mussels a problem in inland waters?

Zebra mussels, especially, are a serious problem in the Great Lakes and other inland waters. Their larvae are carried worldwide in the ballast tanks of ships. When released, the mussels multiply rapidly and may clog the intake pipes of waterworks and power-generating plants. They also foul pumps, navigation buoys, and boat hulls. Their growth upsets food chains and threatens fish populations. There is even the potential of zebra mussels invading plumbing systems and plugging interior sprinkler systems.

2. What does the term *biodegradable* mean?

Such materials readily decompose, usually by bacterial action. Examples include almost all natural materials and some commercial products such as detergents and paper.

3. Who invented superglue?

Superglue has the generic chemical name cyanoacrylate, $C_5H_5NO_2$. It was discovered accidentally in the 1940s by researcher Harry Coover while at Eastman Kodak. The strong, instant glue was popularized on television in 1959 when one drop placed between metal plates was shown to support a person's weight.

OCTOPUS ➜ ROBOTICS

1. Does an octopus regenerate a lost arm?

Yes, octopi can regrow lost arms. Incidentally, these creatures have a short lifetime, usually only six months to five years.

2. How long can octopus arms grow?

The Giant Pacific Octopus has arms that may reach 13 feet (4 m) long. This creature may weigh well over 100 pounds. When newly hatched, however, an octopus may be as small as a grain of rice.

3. What kinds of robotic arms are currently used in space?

Canadian-built robotic arms have been a part of the U.S. space shuttle fleet for decades. The arm, or crane, is 50 feet long and moves successfully with great precision. The International Space Station has a newer generation Canadian robotic arm. This jointed device can move about the station's exterior like an inchworm. That is, the two ends take turns attaching and unfastening from outside brackets on the space station.

SEASHELL ➜ CONSTRUCTION MATERIAL

1. What is the origin of the name *mother-of-pearl*?

The name is centuries old. The word *mother* was used in early English to describe hardened layers, and the word *pearl* may derive from their spherical shape.

2. What materials are currently used for artificial bone?

Traditional implants are made of metal alloys including stainless steel, titanium, and cobalt. New composite materials, suggested by seashells, have many advantages over metals.

3. What are some traditional uses of shell material?

Mother-of-pearl has long been made into buttons by punching disks from seashells. The material is also used in jewelry and is inlaid in furniture and musical instruments.

Sea Slug ➜ Chemicals

1. How many species of sea slug are known to exist?

   The defining limits of sea slugs are not clear, but more than 1,000 species are known. New species are found regularly around the world.

2. What does the term *microbe* refer to?

   The term is applied to any microscopic living thing, either plant or animal, and especially describes bacteria.

3. What are some descriptive sea slug names?

   There are many names, including lettuce, butterfly, angel, and sea cucumber. Some of these colorful creatures are kept in saltwater aquariums.

Sea Sponge ➜ Fiber Optics

1. What is the composition of man-made optical fiber?

   The common material is glass or silica, SiO_2, the composition of sand. Clear plastic strands are also shaped into fiber optics.

2. Have fossil sponges been found?

   Yes, well-preserved sponge fossils are found in rocks around the world. There is evidence that many fossils, including sponges, are a result of the Genesis Flood event.

3. Besides the mutual benefit between the Flower Basket and shrimp, give some other examples of symbiosis.

   There are endless examples of symbiosis in nature. On the microscopic level, nitrogen-fixing bacteria called rhizobia live in the root nodules of legumes. In the sea, clown fish dwell among the tentacles of sea anemones for mutual protection from predators.

Whale ➜ Submarine

1. What is the largest whale?

   Blue whales are by far the largest living animal. They grow to a length of 110 feet (33 m) and weigh more than 200 tons. This is 15 times heavier than the largest elephants. There may have been land dinosaurs that approached the weight of blue whales, but there also may have been larger whales in the past.

2. How deep can whales dive?

   Whales regularly dive, or "sound," to depths of hundreds of feet or meters. Sperm whales have been monitored to a depth of 6,562 feet

(2,000 m), and they may go far deeper. The pressure at this depth is 116 times greater than standard air pressure.

3. How do whales communicate with each other?

Whales make sounds that are variously described as clicks, whistles, cries, howls, and songs. The underwater vibrations allow whales to stay in contact when miles apart. In fact, sensitive instruments can detect whale sounds over a distance of 2,000 miles. The meaning of these whale sounds is poorly understood by researchers.

Chapter 5 — Land Animals

Ankylosaurus ➜ Fiberglass

1. How large were dinosaur eggs?

Fossil dinosaur eggs are either round or oblong, and range from an inch to a foot in size. Nests and petrified eggs are found in many places, including Argentina, Canada, China, India, Mongolia, and western U.S. states.

2. How many dinosaur species have been discovered?

The number is uncertain because many dinosaur finds are only partial fossils. Authorities estimate about 700 known species, and perhaps an equivalent number still unknown.

3. When was fiberglass invented?

Modern fiberglass was invented by Russell Games Slayter of Owens-Corning Company in 1938. Its initial use was for insulation, still a major application today.

Antler ➜ Organ Repair

1. Which animal holds the record size for antlers?

The moose has impressive antlers that may reach six feet (1.8 m) from tip to tip. However, the record antler size belongs to the Irish elk, or giant deer. Now extinct, these majestic animals once lived in Europe. Some fossilized antlers measure 12 feet (3.6 m) across, twice the size of the largest moose antlers.

2. How is antler size related to animal age?

As the animal ages, the antlers grow larger, but there is no direct connection between age and number of "points." A deer's first rack may have two to six points, and the increasing number levels off with age. Antlers are shed each mating season.

3. What are stem cells?

Stem cells occur in all living things, both embryos and adults. They are able to develop into a large number of specialized cells and replicate (duplicate) themselves. Stem cells may become muscle, nerves, organs, or skin tissue. The differentiation of stem cells is controlled by their internal genes.

DOG PAW ➜ SHOE SOLES

1. Where else besides shoe soles is the herringbone pattern found?

The crisscross pattern is found in brickwork, hardwood flooring, paving stones, tire tread, weaving, embroidery, and not the least, in the bone arrangement of herring.

2. How have Paul Sperry's ideas been applied to automobiles?

Many tire treads feature a V-shaped herringbone pattern that channels away surface water while maintaining good traction.

3. Where can one learn more about Paul Sperry's discovery?

Go to sperrytopsider.com.

GECKO ➜ ADHESIVE

1. Explain the nature of van der Waals and capillary forces.

J. D. van der Waals (1837–1923) was a Dutch physicist. The force named for him is a weak electrical attraction between molecules. The capillary force is a similar weak molecular attraction that draws water up a straw, or into a paper towel.

2. Who invented Post-it notes?

Two researchers at 3M Company, Spencer Silver and Arthur Fry, made the weak adhesive in the laboratory. Arthur later realized that this material could hold temporary bookmarks in his church choir hymnal. In 1977, 3M began marketing the popular post-it notes.

3. How do spiders and other insects walk on ceilings?

Many of them have microscopic hairs on their legs that function similar to geckos.

GIRAFFE ➜ ANTIGRAVITY SPACESUIT

1. What exercise activity of orbiting astronauts counters "fluid shift"?

There are a number of space hazards besides fluid shift, including muscle weakening and bone density loss. Astronauts perform daily exercises to counter these effects and maintain health. These include treadmills and bungee-type stretching. Efforts to control fluid shift include the wearing of constrictive cuffs.

2. How tall is an adult giraffe?

    Male giraffes grow somewhat taller than females, reaching a height of 16–20 feet (5–6 m). The neck and shoulder height averages 12 feet (3.7 m). The giraffe has seven neck vertebrate, the same as people.

3. How large is the giraffe heart?

    To provide sufficient blood pressure, the heart of an adult giraffe weighs about 22 pounds (10 kg). This is 30–40 times heavier than a human heart.

HORSE BONE ➜ CONSTRUCTION

1. Do we have any bones with a foramen opening, similar to the horse?

    Yes, there are natural openings or foramen in some of our bones, for example, our upper arm humerus. Beside horses, other animals also have foramen bone openings for the entrance of nerves and blood vessels.

2. What is the top speed of a race horse?

    Race horses achieve speeds approaching 40 miles per hour.

3. How much stress is generated in the bones of athletes?

    When a person stands on one foot, the stress in the tibia leg bone can reach the extremely high value of 1,450 pounds/in^2 (10^7 newtons/meter^2). For its protection from fracture, the bone slightly contracts when under such stress.

PENGUIN EYE ➜ SUNGLASSES

1. Where in the world are wild penguins found?

    Penguins inhabit many parts of the southern hemisphere. This includes Antarctica and the southern tips of continents. One penguin species is found in the Galapagos Islands, on the equator.

2. How large are penguins?

    The height of emperor penguins reaches 3.5 feet (1.1 m). In contrast, the adult fairy penguin is only 1.5 feet tall (0.5 m).

3. What is it about the color orange that makes a useful light filter?

    Orange-tinted sunglasses reduce or eliminate blue light, a major component of glare. Ultraviolet light is also at the blue end of the spectrum, and "blue-blocking" sunglasses minimize eye damage from UV.

Tree Frog → Automobile Tires

1. In what regions do tree frogs live?

   Tree frogs are common across mid-North America. They also range into Asia and North Africa.

2. How do tree frogs act as barometers?

   These creatures become very noisy on spring evenings, especially when rain is approaching. A falling barometer precedes rain, and tree frogs are somehow sensitive to the changing air pressure. The decibel level of a group of tree frogs can reach an impressive 70 decibels, rivaling the noise of a lawn mower or chainsaw.

3. How many species of frogs are known in the world?

   There are more than 5,000 frog species. Frog populations have declined in recent decades and the reason is uncertain. Suggestions include climate change, disease, habitat loss, pollutants, and predators.

Chapter 6 — People

Body Odor → Insect Repellant

1. Which mosquitoes bite, the males or females?

   It is only the females that bite, in order to get a blood meal. Protein obtained from the blood helps their egg production. Both male and female mosquitoes are also nectar feeders.

2. What chemicals are typically used in insect repellants?

   There are a large number of chemical repellants, including citronella, permethin, picaridin, DDT, and DEET. Also somewhat effective are natural oils from catnip, eucalyptus, and soy beans.

3. What is a major worldwide danger of mosquitoes?

   Mosquitoes are vector agents. That is, they carry harmful viruses and parasites from person to person without getting sick themselves. Especially dangerous are yellow fever, dengue fever, and malaria. Millions die from malaria each year, many of them African children.

DNA → Computer Memory

1. What does DNA stand for?

The letters are shorthand for deoxyribonucleic acid. This is a twisted or helical chain structure present within all living cells. Researchers James Watson and Francis Crick are credited with the discovery of DNA's helical form in 1953.

2. What is the size of a DNA molecule?

The tightly wound helical structure is about 25 angstroms wide and millions of times longer. If uncoiled, a DNA molecule would stretch to about six feet. It would be invisible since it would consist of a string of individual atoms.

3. What do the letters stand for in Einstein's formula $E = mc^2$?

This equation expresses the equivalency of energy (E) and mass or matter (m), related by the speed of light (c). One gram of matter, if completely converted to energy, would produce the equivalent of 20,000 tons of explosive. Providentially, the energy release process is difficult.

EARDRUM → EARPHONE

1. What were some of Alexander Graham Bell's other inventions?

Bell shared in 30 patents covering a vast range of inventions. They include the phonograph, nickel-iron battery, audiometer, metal detector, and vehicles for air and water travel.

2. What are the names of the three bones in our inner ear?

These are the smallest bones in the human body. Based on shape, they are called the hammer, the anvil, and the stirrup. Their Latin names are the malleus, incus, and stapes.

3. How was Alexander Graham Bell honored at his death on August 14, 1922?

On this day, all telephones served by the Bell Telephone system went silent for one minute as a tribute to the great inventor. In his later years, Bell did not like interruptions and refused to have a telephone placed in his study.

EYE IRIS → IDENTIFICATION

1. How does the iris control incoming light to the eye?

The iris acts like the aperture or diaphragm of a camera, controlling the amount of incoming light. The iris consists of a colored fibrovascular film of cells. Depending on the amount of light available, muscles cause the iris to either dilate or constrict. This expansion and contraction controls the light entering the interior of the eye through the pupil. In darkness, the iris pulls back and the pupil opens wide.

2. What are some ways to express the number 10^{78}?

This number, one followed by 78 zeroes, is entirely beyond useful comparisons. The known stars roughly match an estimate of all the sand grains on the seashores of earth, 10^{22}, or ten billion trillion. This number is 56 zeros smaller than 10^{78}. The number of atoms in all the visible stars of the universe is estimated at 10^{70}. This immense number is still 100 billion times smaller than 10^{78}.

3. Where is iris recognition currently used for security?

Iris technology is used at several airports and international border crossings. Some U.S. public schools and businesses are also experimenting with the identification method.

FIBRIN ➜ ELASTIC

1. What is hemophilia?

Hemophilia is a genetic disorder in which certain blood clotting factors, or chemicals, are absent. As a result, the body cannot make fibrin to seal wounds and bleeding persists for a long period of time.

2. How does Coumadin affect blood clotting?

Coumadin is a natural anticoagulant found in many plants. The artificially formed variety is called warfarin. It inhibits the formation of vitamin K clotting factors. This and similar medicines are called vitamin K antagonists, and are effective in thinning the blood and preventing strokes.

3. How is the blood clotting mechanism related to intelligent design?

The mechanism of blood clotting requires precise reactions of many interrelated chemicals. If any one of these biochemical steps is missing, the mechanism may fail. An entire graduate course could be built on our limited understanding of the blood clotting process. Many scientists conclude that blood clotting is far too complex, and essential, to evolve by mere chance. Instead, it shows divine planning of the internal workings of living creatures.

FINGERPRINT ➜ PROSTHETIC HAND

1. Is every fingerprint unique?

Yes, fingerprints are unique, even for identical twins. In such cases there are close similarities, but also measureable differences.

2. What typical frequencies do our fingerprints detect?

From the spacing of fingertip ridges and typical motions of the hand, the resulting vibration passed on to sensory nerves centers around 250 cycles/second.

3. Do primates have fingerprints similar to people?

There is great variety of fingertip detail in the animal world. The most widespread monkeys, called macaques, have straight-line ridges parallel to their fingers, rather than our familiar swirls. This is one of many differences between people and animals.

LEG BONE ➔ EIFFEL TOWER

1. What is the largest bone in the human body?

The femur, from the Latin word for thigh, is our longest and strongest bone. The average length for adults is 17 inches (43 cm).

2. How much does the Eiffel Tower weigh?

The tower weighs about 10,000 tons. It is an open and relatively lightweight structure. A large cylinder of air, the size of the tower at its base and the same height, would weigh more than the tower itself.

3. How many people visit the Eiffel Tower each year?

About seven million people visit the Eiffel Tower in Paris each year, the most visited paid monument in the world.

MUSCLES ➔ ROBOTICS

1. How many muscles are in the human body?

The number varies between 650 and 850, depending on how muscle bundles are distinguished and counted. The arm contains dozens of distinct muscles.

2. Which metals have the largest thermal expansion?

Some common metals with high thermal expansions are aluminum, magnesium, selenium, tin, and zinc. Lengths of these metals increase by 10-20 millionths of the original length, with each degree of temperature increase.

3. How much weight is an adult capable of lifting?

Consider the "clean and jerk" event in which the barbell is first lifted to a front squat, then jerked overhead in a standing position. The record lift is 586.4 pounds (266 kg). The muscles of the human body display very impressive strength.

Saliva ➜ Healing

1. Would not saliva infect an open wound?

   Just the opposite. Saliva contains several compounds that are antibacterial in nature, and they hasten healing.

2. Why do animals sometimes lick their wounds?

   This common practice is a form of self-medication called *zoopharmacognosy*. Similar to people, animal saliva has healing qualities. Animals are "programmed" by their Creator to cleanse and treat wounds as best they can.

3. How was the healing property of histatin verified?

   Scientists cultured skin cells in a dish, then scratched them and treated some with saliva. The "wounds" of the treated cells healed much more quickly than the untreated cells.

Skin ➜ Self-repairing Plastic

1. How many square feet of skin does an adult have?

   Our skin averages 14–21 square feet (1.3–2 m^2) of surface and comprises 15 percent of body weight. Just one square inch of skin typically has 20 blood vessels, 650 sweat glands, and more than 1,000 nerve endings.

2. What is plastic surgery?

   The word *plastic* comes from the Greek verb meaning "to mold or shape." Plastic surgeons usually do not use synthetic polymer or plastic. Instead, they reshape a person's bone, cartilage, muscles, fat, or skin. The purpose may be cosmetic or the repair of injury.

3. Is artificial skin available in medicine?

   There is much current research on the health needs of burn victims. Sheets of artificial skin are available to cover exposed areas and allow the lower, dermal skin layer to repair. At a later time, the synthetic layer is removed, allowing the outer, epidermal skin layer to grow. The sheets are variously made of animal collagen, silicon, or nylon. There is also some success with cultured skin, taken from the patient and grown externally, then later reapplied.

Tooth Enamel ➜ Armor Coating

1. What is the composition of tooth enamel?

The tooth coating is chiefly the mineral hydroxyapatite. It is also called calcium phosphate with formula $Ca_{10}(PO_4)_6 \cdot 2(OH)$. This chemical is also one component of our bones.

2. Did George Washington have wooden false teeth?

   This is a popular story that is not true. Washington lost his teeth at an early age due to gum disease. He had several dentures prepared over the years, some of which are on display at the Smithsonian Museum. One set is made of ivory from the tusk of a hippopotamus.

3. What material is used in making modern artificial teeth?

   Many centuries ago, false teeth were made of metals such as gold or lead, and also from human or animal teeth. In colonial times, bone and ivory were popular denture materials. Modern dentures consist of plastic, porcelain, or high-quality acrylic resin. The latter artificial teeth are molded and tinted to match each patient.

Vernix → Skin Cream

1. What is the origin of the name *vernix caseosa*?

   Vernix is Latin for "varnish," describing the coating on the newborn. The second word is Latin for "cheesy," also describing the appearance of the protective material.

2. Do animals also produce vernix?

   No other land mammals, including apes, produce vernix-type material in the fetal stage. This is a problem for evolutionary linkages between animals and people. Some sea mammals do produce an equivalent birth material that is valuable to babies in the watery world.

3. What is ultraviolet light?

   Ultraviolet light, or UV, is also called black light. It has a wavelength shorter than visible light, and causes sunburn on unprotected skin.

Chapter 7 — Vegetation

Beech Leaf → Space Antenna

1. What is the origin of the word *origami*?

   Origami is the ancient Asian art of constructing intricate objects and animals with folds and creases in paper. The word has Japanese roots that mean "paper folding."

2. How does a butterfly unfold its wings?

When a butterfly emerges from the pupa, its wings are crinkled and wet. The initial, intricate folded nature of the wings has not yet been analyzed. The butterfly pumps blood into the wing veins to inflate them. As the wings dry, they stiffen and become useful for flight and also oxygen exchange.

3. Is there an example building structure based on origami?

One example is a folded bamboo house used for emergency shelters. It was designed by Tang & Yang Architects of Savannah, Georgia.

CHEMICALS → MEDICINE

1. How was aspirin discovered?

The chemical called aspirin forms naturally in the bark of the weeping willow tree. The use of willow bark to reduce fever is ancient, and is described in fifth century B.C. writings from Hippocrates. Native American Indians also discovered the pain-relieving nature of aspirin and made medicinal tea from willow tree bark. The drug was first synthesized by German chemist Hermann Kolbe in 1859.

2. What actually is aspirin?

The name is short for acetylsalicylic acid, or salicylic acid. The chemical formula is $C_9H_8O_4$. This white, crystalline drug is useful for relieving a number of ailments.

3. Does Scripture comment on plant use for healing?

Revelation 22:2 describes future trees in the New Jerusalem. Their leaves are useful for "the healing of the nations." Ezekiel 47:12 further describes these medicinal leaves.

COCKLEBUR → VELCRO

1. What are some novel uses of Velcro?

Beyond clothing, some uses of Velcro include blood pressure cuffs, car bumpers where the hooks and loops are made of stainless steel, and ready-to-assemble furniture.

2. How strong is Velcro?

On flexible surfaces, Velcro can be readily pulled loose. On rigid surfaces, where all the hooks and loops must separate at once, the strength is far greater. Depending on the surface area, the holding force may be hundreds of pounds.

3. Where is Velcro Valley?

The location is an industrial region of Orange County, California, where hundreds of companies make sports apparel. The name comes from the instant success of startup companies with new ideas.

Fava Bean ➜ Valve

1. What is the origin of the name *fava*?

The word is Italian for bean. Further names for this legume include bell, broad, field, tic, and vicia faba.

2. What are fava beans used for today?

Some countries cultivate fava beans for food. The plant also has a root system that prevents erosion and absorbs nitrogen from the soil.

3. What is an unusual use for fava beans?

In ancient Greece and Rome, such beans were used in voting. A white bean counted as a yes and a black bean meant no.

Fescue Grass ➜ Herbicide

1. What is the most-used herbicide worldwide?

The herbicide glyphosate is sold under such brand names as Roundup and Touchdown Total. Annual worldwide use is estimated at 15–20 million pounds in agriculture, and about one-third this amount additionally in landscaping.

2. What are some other examples of allelopathy?

Allelopathic vegetation includes spotted knapweed, garlic mustard, nutsedge, black walnut, eucalyptus leaf litter, and several desert shrubs.

3. What are some varieties of fescue grass?

More than 300 distinct species of fescue are catalogued. They vary in height from just inches to over six feet (several cm to 1.8 m).

Lotus Flower ➜ Surface Cleaner

1. What is the origin of the name *sacred lotus*?

The beauty of the flower has long been associated with divinity. The plant was venerated in early Asia, Egypt, and Persia. The water lily remains the national flower of India and Vietnam. Along with its beauty, all parts of the plant are also edible.

2. What does the seed head of the water lily resemble?

The dried disk resembles the sprinkler spout of a watering can. These seed pods are often used for tabletop decoration when dried.

3. What is surface tension?

This is the tendency of liquid molecules to adhere together. Two liquids with extremely high surface tension are the element mercury and water. Both readily form beads on a smooth surface. This property of water is very important. As two examples, the surface tension keeps our joints lubricated, and also allows water to move to the top of trees through tiny capillary tubes within the trunk.

Osage Orange → Barbed Wire

1. Does Osage Orange produce edible fruit?

The bush produces a round, bumpy fruit four to five inches in diameter. It is not edible for people or animals. However, its oils are found to repel pests, including cockroaches, mosquitoes, and spiders.

2. Where might one find a "barbed wire museum"?

Sites include the the Devil's Rope Museum in McLean, Texas, Ellwood House and Museum in DeKalb, Illinois, and the Barbed Wire Museum in LaCrosse, Kansas. The facilities display over 2,000 varieties of barbed wire.

3. What is razor wire?

This is a registered trademark name. The barbed wire or "tape" is a mesh of sharp metal strips designed to prevent passage by people. It is an effective visual deterrent used where high security is needed.

Pine Cone → Smart Clothes

1. Are some pine cones opened only by fire?

Yes, the cones of come jack pines and California Torrey pines may remain dormant for 10–20 years. The eventual, intense heat from a forest fire causes them to open and disperse their seeds.

2. Why do some pine cones open up when warm and dry, and close again when wet and cold?

It is female pine cones that open when warm and dry, the best conditions for germinating their seeds by wind-blown pollen from male pine cones.

3. What are some future plans for smart clothes?

Future "healthy clothes" may have built-in sensors to monitor the wearer's heart rate, breathing, and temperature.

RUBBER TREE ➜ AUTOMOBILE TIRES

1. How is tree gum obtained from the rubber tree?

   The white sap-like gum, called latex, occurs in capillary tubes that spiral up the tree within the bark. These vessels are cut and drained without harming the tree, somewhat like the tapping of maple trees.

2. Besides pencil erasure, what other great discovery was made by Joseph Priestley?

   In 1774 Priestley isolated and identified oxygen gas. Oxygen makes up 21 percent of our atmosphere, next after nitrogen, which comprises 79 percent.

3. How does sulfur benefit the durability of rubber?

   Rubber consists of long polymer chains, which are molecules with repeated units of carbon and hydrogen atoms. Sulfur atoms provide strong cross-link connections between the chains, tying them together. The result is rubber, which is harder, more durable, and resistant to chemical breakdown.

SKUNK CABBAGE ➜ THERMOSTAT

1. Describe the odor of the skunk cabbage.

   The odor of an undisturbed plant is mild. If a leaf is damaged, however, a definite skunk-like odor results. This smell attracts insects for pollination, and also discourages eating by larger animals.

2. Besides the skunk cabbage, what other plant is known for its offensive odor?

   There are many flowers that do not "smell like a rose." The Arum family of flowers, with about 25 species, is also known as a corpse flower. The champion may be titan arum, which grows in Indonesia. Its enormous blossom may reach several feet across. The smell, something like rotting flesh, has been known to make people pass out. Certain insects are drawn to this odor and serve as pollinators.

3. How does a thermostat bimetallic strip function?

   Two distinct metals are fused together. These metals, copper and aluminum, for example, have different coefficients of thermal expansion. This means that they expand differently with increasing temperature. This causes the metal strip to bend or curl as temperature changes. The resulting movement allows electrical contact, which turns a circuit either on or off.

Spinach ➜ Solar Cell

1. What is the chemical equation for photosynthesis?

The following equation describes the conversion of light energy to chemical energy by living organisms:

$$6CO_2 \text{ (gas)} + 6H_2O \text{ (liquid)} + \text{light energy} \rightarrow$$
$$C_6H_{12}O_6 \text{ (glucose)} + 6O_2 \text{ (gas)}$$

2. How do silicon solar cells work?

Light photons, or particles, give their energy to outer electrons of carbon atoms. The electrons that are knocked loose, and the resulting "holes" left behind, can then drift through the silicon crystal, resulting in an electric current.

3. What is the potential of solar energy on earth?

Every hour, solar energy hitting the earth supplies more energy than the entire world population uses in a year. Much of this energy is reflected back into space or absorbed by the oceans. Still, there is great potential for meeting earth's energy needs by harnessing solar energy. Current challenges include the low efficiency of present technology. Today, only about 0.1 percent of the world's electric energy needs are met by solar collectors.

Venus Flytrap ➜ Food Packaging

1. Where do Venus flytrap plants grow?

The plants are not tropical and can survive mild winters. Many are found growing wild in North and South Carolina and Florida. They prefer the nitrogen-poor soil found in bogs. As a houseplant, the Venus flytrap requires high humidity, as in a terrarium.

2. What other plants are carnivorous, or insectivorous, besides the Venus flytrap?

Pitcher and cobra plants trap prey in a liquid chemical pool held within a rolled leaf. The sundew plant attracts insects to a sticky surface, like flypaper. Bladderworts suck aquatic insects into their interior. Over 1,000 carnivorous plant species are known.

3. Do flytraps ever catch animals larger than insects?

Flytraps can catch large beetles, grasshoppers, and small frogs or mice. However, they are no danger to larger pets!

WATER LILY ➔ CONSTRUCTION

1. What is the origin of the name *Victoria amazonica*?

   The flower is named for England's Queen Victoria (1819–1901). She was British royalty during the time of the Great Exhibition in 1851. The water lily lives in the shallow tributaries of Brazil's Amazon River.

2. What was inside the Crystal Palace?

   The building held 14,000 technology displays gathered from around the world. This period, part of the Victorian era, was a high point of the industrial revolution. Featured at the exhibition were steam power, electricity, and other new technologies. Popular features in the Crystal Palace included elegant restrooms, called "retiring rooms."

3. What is the water lily vein pattern?

   It is called a symmetric radial dendrite pattern. From the center of the lily pad, smaller veins branch off from larger veins.

WILD WHEAT ➔ HUMIDITY SENSOR

1. How do wild and domestic wheat differ?

   One major difference is that wild wheat readily sheds its ripened grain. This blowing seed then plants itself and grows elsewhere. With domestic or cultivated wheat, the grain tightly adheres to the dead stalk. This retention allows easy harvesting.

2. What are some common names for wheat varieties?

   The host of adjectives for wheat include winter, miracle, soft (high in starch), hard (high in protein), and common wheat, often used for bread flour.

3. Where is the filaree plant found?

   This plant is abundant in western United States and Canada. It prefers dry, sandy soil.

Chapter 8 — Nonliving Objects

BUCKYBALLS ➔ MICRO BALL BEARINGS

1. What is the geometric shape on the surface of soccer balls and buckyballs?

   There are two interconnected shapes, pentagons (five sides) and hexagons (six sides). Both buckyballs and soccer balls have surfaces covered with 12 pentagons and 20 hexagons.

2. How does the size of a buckyball compare with the period at the end of a sentence?

   Roughly one quarter million (250,000) buckyballs could fit side-by-side across the width of a period.

3. What is the benefit of an architectural geodesic dome?

   Such domes, often made largely of glass, have great strength because the supports are under compressional forces. The buckyball is similarly strong because of the many chemical covalent bonds between carbon atoms.

Nanoparticles → Water Purifier

1. Which countries are in special need of clean water?

   Water pollution is a worldwide problem. It is especially serious in Africa, Asia, and East Europe.

2. Do some bacteria "eat" oil spills?

   Yes, there are microorganisms that decompose oil. The bacteria use enzymes to break down the oil molecules, and then consume them for energy. This happens continually across the earth at sites of natural oil seepages and man-made spills.

3. One nano iron particle, the size of the finest dust, contains how many iron atoms?

   Such a particle would contain a thousand or more iron atoms.

Opals → Photonic Materials

1. Where are natural opals found?

   Leading countries for opal output are Australia, Ethiopia, Mexico, and the United States About 97 percent of new opal comes from Australia.

2. What is lithography?

   Lithography is a printing process using chemicals to form an image. Today it includes the production of integrated electronic circuits.

3. How can a computer operate using light pulses?

   Instead of moving electrons, light beams are used. Research continues to replace electrical transistors, switches, and gates with optical analogs.

Pulsar ➜ Time Standard

1. What is the typical density of a neutron star?

   Such stars are incredibly dense. Their mass or weight is comparable to the sun, yet they are 150,000 times smaller. Just one teaspoon of neutron "stardust" would weigh a billion tons.

2. How close to earth is the nearest known pulsar?

   The nearest known neutron star is in the direction of the southern constellation Corona Australis, 200 light years away.

3. What would happen if we visited a neutron star?

   Such a visit would be perilous. Near the surface we would be vaporized by the star's intense heat and radiation. In addition, with a gravity force of 100 billion times that of earth, any nearby object would be flattened out to a single layer of atoms.

Water Flow ➜ Impeller

1. Is there a mathematical formula for the spiral designs found in nature?

   A number of precise, elegant equations describe the gentle curves of spirals. Archimedes' spiral has the polar equation $r = a\theta$. A logarithmic spiral is given by $r = ae^{b\theta}$. In these formulas r is radius distance, θ is angle; a and b are constants.

2. How does an impeller differ from a propeller?

   Both items have rotating blades that move air or liquids. An aircraft or boat propeller thrusts air or water in a desired direction, and the vehicle recoils in the opposite direction. An impeller is usually placed in a pipeline to move a confined fluid.

3. What is the largest spiral observed in nature?

   This honor goes to spiral galaxies, such as our own Milky Way. Such galaxies are about 100,000 light years in diameter, or 600 thousand trillion miles (6×10^{17} miles), and contain 100 billion stars each.

GENERAL REFERENCES

Benyus, Janine. 1997. *Biomimicry: Innovation Inspired by Nature.* Harper Collins Publishers, New York.

The Biomimicry Institute. This secular organization promotes the transfer of ideas from nature to solve human problems. BiomimicryInstitute.org

Forbes, Peter. 2005. *The Gecko's Foot.* W.W. Norton & Company, New York.

Frenay, Robert. 2006. *Pulse.* Farrar, Straus and Giroux, New York.

Hutchins, Ross E. 1980. *Nature Invented It First.* Dodd, Mead, and Company, New York.

Laithwaite, Eric. 1994. *An Inventor in the Garden of Eden.* Cambridge University Press, Cambridge.

Paturi, F.R. 1976. *Nature, Mother of Invention: The Engineering of Plant Life.* Thames and Hudson, London.

Peterson, Roger Tory, Editor. Peterson Field Guide Series. *Medicinal Plants and Herbs of Eastern and Central North America* (Steve Foster and James Duke, 1999); *Western Medicinal Plants and Herbs* (Steve Foster and Christopher Hobbs, 2002). Houghton Mifflin Company, New York.

Sarfati, Jonathan. 2008. *By Design.* Creation Book Publishers, Powder Springs, GA.

Vogel, Steven.1998. *Cat's Paws and Catapults.* W.W. Norton and Company, New York.

Willis, Delta. 1995. *The Sand Dollar and the Slide Rule.* Addison-Wesley, Reading, MA.

GLOSSARY

bioluminscence

The generation of light by living organisms using chemical reactions. Bioluminscence is found in many species of bacteria, algae, fungi, invertebrates, insects, and deep-sea fish.

biomimicry (also called biomimickry, biomimetics, bionics, biognosis, and bio-inspiration)

The use of ideas from nature to develop new products and solve problems. The word *biomimicry* comes from two Greek words meaning life (*bios*) and imitation (*mimesis*).

bionomics

The use of biological principles in nature to model and analyze economic trends.

calcite

Also called calcium carbonate, or limestone, with the formula $CaCO_3$. This strong mineral material is frequently found in shells, bone, and the eye lenses of some marine animals.

chitin (kI-tin)

The protective covering of arthropods, including beetles. Chitin consists of nitrogen-containing molecules called polysaccharides.

composite material

Two or more combined substances with beneficial properties beyond themselves alone, usually greater strength.

echolocation

The use of reflected sound waves to determine the distance and direction of nearby objects.

ethnobotany

The study of native societies to learn their use of plants for dyes, food, medicine, etc.

extremophiles

Plants or animals that survive climate extremes such as intense heat, cold, saltiness, or drought.

fiber optics

Flexible strands of clear glass or plastic that can carry digital light signals over long distances.

keratin

A protein commonly found in the beaks of birds, feathers, hair, and horns. Keratin also composes our fingernails.

mems

An acronym for microelectromechanical systems. These tiny robotic devices range in size from microns to centimeters.

micron

A length equal to one millionth of a meter, or 0.00004 inch. The thickness of a human hair is about 100 microns.

mother-of-pearl

See nacre.

nacre (NAY-Ker)

A tough layering of calcite ($CaCO_3$) in a protein matrix. Also called *mother-of-pearl*, it coats many seashells. Nacre is many hundreds of times stronger than chalk, another common form of $CaCO_3$.

nanofabrication

The building of materials upward from the scale of molecules, usually layer upon layer.

nanometer

A length of one billionth meter (4×10^{-7} inch). The diameter of an atom is about one-tenth nanometer. There are about 100,000 nanometers in the thickness of this page of paper, or in the width of a human hair. The word *nanos* is Greek for "dwarf."

nanotechnology

Technology on the very small scale of nanometers. One hundred nanometers is a thousand times smaller than a human hair. The journal titled *Nanotechnology* is dedicated to understanding and developing the small scale, including bearings, motors, and optical devices.

polymer

A compound made up of many smaller molecules, called monomers, which are linked together. Cellulose is a natural polymer, while nylon and plastic are synthetics.

synthetic biology

The effort to modify plants and animals, producing new capabilities in them.

zoopharmacognosy

The study of the use of plants and earth minerals by animals for their self-medication.

SUBJECT INDEX

232

DON DEYOUNG

Don DeYoung chairs the Physical Science Department at Grace College, Winona Lake, Indiana. He holds a doctorate in physics from Iowa State University and a Master of Divinity from Grace Seminary. He has written fifteen books on Bible-science topics including astronomy, pioneer creation scientists, and object lessons for children.

Dr. DeYoung is currently president of the Creation Research Society with 1,700 members worldwide. He is an active speaker on creation topics and believes that the details of nature are a powerful testimony to the Creator's care for mankind. Dr. DeYoung and his wife Sally have three married daughters.

DERRIK HOBBS

Derrik Hobbs comes from the business world, providing market strategies to farming operations, individuals, and corporations that have risk exposure to commodities. He has also been an investment advisor for professional athletes, corporate executives, and small business owners. He is the author of a book on technical analysis of the stock and commodity markets.

Derrik has an active interest in creation studies including business models based on principles and processes found in nature. Derrik and his wife Jessica have three daughters. He is a son-in-law of coauthor Don DeYoung

Thousands . . . not Billions

Challenging an Icon of Evolution,
Questioning the Age of the Earth

Dr. Don DeYoung

Eight respected Ph.D.s expose the truth about radiometric dating and years of science fiction!

This long-awaited book shatters the famed dating methods employed by evolutionists to cast doubt on the veracity of the Bible and its chronology of earth history.

Radiometric dating is one of the linchpins of evolutionary education today. By dating the soil in which fossils are found to very long ages, evolutionists undermine faith in Genesis as the true documentary of the history of the universe. When people are told that a dinosaur bone has been determined to be tens of millions of years old, that obviously doesn't square with the biblical record of man being created on day 6 with the land animals.

But DeYoung now demonstrates that Christians no longer have to puzzle over this glaring contradiction. A must-have for the serious Bible student, *Thousands . . . not Billions* will bolster the faith of many.

6 x 9 • paperback • 192 pages • $13.99

ISBN-13: 978-0-89051-441-2 • ISBN-10: 0-89051-441-0

Also available in a 2-DVD set • ISBN: 827087091192 • $19.99

Available at Christian bookstores nationwide or go to www.nlpg.com

Frozen in Time
Woolly Mammoths, the Ice Age, and the Biblical Key to Their Secrets
Michael Oard

What secrets are revealed in the frozen woolly mammoth "time capsules"? And how can the Bible unlock their secrets in ways that evolution-based science cannot?

The Ice Age is one of the most difficult eras in geological history for a uniformitarian (those who believe the earth evolved by "slow processes over millions of years") scientist to explain, simply because long ages of evolution cannot explain it. Many mysterious questions about the Ice Age arise:

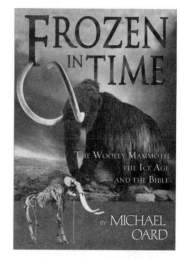

- What would cause the summer temperatures of the northern United States and Europe to plummet over 50 degrees Fahrenheit?
- What was the source of the abnormal amount of moisture necessary for heavy snow?
- What caused the cold summer temperatures and heavy snowfall to persist for hundreds of years?
- Why did mammoths become extinct, not only in Siberia, but also across the earth, and at the same time as many other large mammals?
- How could their remains still have partially decayed food in their stomachs?

Author Michael Oard gives plausible explanations of the seemingly unsolvable mysteries about the Ice Age and the woolly mammoths in this book. Many other Ice Age topics are explained, including super Ice Age floods, ice cores, man in the Ice Age, and the number of ice ages.

6 x 9 • paperback • 224 pages • $13.99
ISBN-13: 978-0-89051-418-4 • ISBN-10: 0-89051-418-6

Available at Christian bookstores nationwide or go to www.nlpg.com

MASTER BOOKS®
SCHOLARSHIP
ESSAY CONTEST

$3000 COLLEGE SCHOLARSHIP

The Master Books® $3000 college scholarship is open to any high school junior or high school senior or the equivalent thereof from any public, private, or homeschool venue. The applicant must be a U.S. citizen and have a minimum GPA of 3.0 or above (on a 4.0 scale). This scholarship is a one-time award and may be used at any accredited two-year, four-year, or trade school within the contiguous United States. This award covers only tuition and university-provided room and board. The scholarship monies will be forwarded to the college, university, or trade school of the winner's choice upon receipt of a copy of the winner's confirmed admission to their chosen school.

Visit www.masterbooks.net to download your application along with:

 Essay Topic and Deadlines

 Rules, Regulations, and Conditions of Eligibility

Master Books — your Creation Resource Publisher — is a division of New Leaf Publishing Group.